SCIENCE OF COMMAND AND CONTROL:

Part II

COPING WITH COMPLEXITY

Edited by
Stuart E. Johnson and Alexander H. Levis

Published in cooperation with the National Defense University

AFCEA International Press
Fairfax, Virginia

Library of Congress Cataloging-in-Publication Data

Command and Control Research Symposium (2nd: 1988: Naval Postgraduate School)
    Science of command and control.

    (AIP information systems series; v. 2)
    "Revised and updated versions of papers presented at the Joint Directors of Laboratories' Command and Control Research Symposium of June 1988"—T.p. verso.
    "Published in cooperation with the National Defense University."
    Bibliography: p.
    1. Command and control systems—Congresses.
2. Military art and science—Decision making—Congresses. 3. Command and control systems—United States—Congresses. I. Johnson, Stuart E., 1944- . II. Levis, Alexander H. III. National Defense University. IV. Title. V. Series.
UB212.C64 1988   355.3'3041   89-14915
ISBN 0-916159-18-3

Table of Contents

Introduction

No area of national defense is more fraught with complexity and paradox than the field of command and control. Its complexity results not only from the sophistication of the technology, but also from the magnitude of its task: the effective employment of the enormously complex and diverse national security apparatus we have acquired since World War II. The paradox is that our great strengths in technology have not translated into a corresponding strength in effective command and control.

Command and control is the meeting point of the grand visions of U.S. national security policy and the dirty business of going to war. So, it is by its very nature, complex.

Why have we been so severely inhibited in realizing the full potential of our technology in this field? This is the question that a group of researchers and practitioners in the field of command and control struggle with year in and year out and who gather in June at the annual Command and Control Research Symposium to share insights: the best of which are presented in this volume.

Beyond the complexity of the technology that compresses the physical building blocks of command and control, there is a schism between the acquisition and employment communities that makes it hard to discuss the matter. There is no proper, common, playing field.

Like it or not, in the "Pentagon's Planning, Programming, and Budgeting System," command and control capabilities are defined and discussed almost exclusively in terms of hardware systems. Questions of the implications of a new technology for military doctrine, organization, and procedure are considered, if at all, on a fragmented, "post facto" basis. This often condemns us to applying new technologies to old problems rather than permitting new technologies to ride new concepts tailored to their potentials, to redefine or even obviate the traditional problems and threats.

We speak of a Western lead in the technologies relevant to command and control. They are real and they are significant. Dr. Phillip Hanson of the University of Birmingham in the U.K. estimates them to be:
- microelectronics: 8-12 years
- general use computers: 8-12 years
- software: 10-11 years

Exploitation of the potentials of these technologies requires that we in the defense community step back and think differently. The research community, whose work is represented here, searches for those human and organizational factors that make it possible to use new technologies to realize the excellence in command and control that is so vital in modern warfare.

For example, we seek to use superior command and control to manage the tempo of the battle, to disrupt enemy actions, and to assess a complex situation and strike the critical (not necessarily the most) targets at the right time. This is an issue that transcends technology and requires thoughtful research and analysis to determine how to put this capability into the battlefield.

In the first volume of this series, emphasis was placed on the role of C^3 systems in helping the commander cope with uncertainty. The underlying notion in the design of many C^3 systems is that timely and accurate data, provided to the commander in a clear and concise manner, will reduce some of the "fog of war" and lead to better, rational decision making. In a sense, this focus stresses the situation assessment aspect of the command and control process. It highlights such technological problems as detection and surveillance, distributed sensing, data fusion, displays, and decision aiding.

There is a second, but equally important focal point for evaluating command and control. Consider the commander who has assessed the situation, developed a plan, and made a decision. How well does the C^3 system support him in disseminating his decision and have it implemented in a timely manner by his forces? It is here, more than anywhere else, that the issue of complexity arises and the need to cope with it becomes acute.

Two of the sources of complexity are: diversity of mission and resources, and advanced technology. Both are closely interrelated and feed on each other. A modern force must be prepared to handle, with little advance notice, a wide variety of missions ranging from low intensity, highly localized, protracted hostilities to global warfare. The scale may involve only a specified command or it may involve multinational forces of a regional alliance. This diversity introduces organizational complexity of the type that is neither well-understood nor is it amenable to stock solutions. Is it preferable to employ highly centralized, multilevel organizational forms that do large scale resource allocation, or is it better to have a much more flat organizational structure? While doctrine specifies centralized control and decentralized execution, the problem remains of establishing the parameters that specify the degree of centralization and the scale at which it is applied. The vexing aspect is that theory, as well as practice, shows that there is no general answer and that different circumstances require different organizational forms for successful execution of a mission. This leads naturally to the notion of flexible, reconfigurable systems; such a notion adds another layer of complexity, that of managing the flexibility.

This brings us to the second source of complexity: advanced technology itself. Multipurpose platforms equipped with a wide variety of sensor and weapon systems, each with a broad set of possibly overlapping capabilities, create an enormously difficult and very real coordination problem. For example, the effective management of EW resources in an air/land battle is a non-trivial matter characterized by a coordination problem of extraordinary complexity. The C^3 system has the task of affecting the necessary coordination.

Thus, advanced technology has expanded the capabilities and potential effectiveness of our forces, but it has also made possible a much more complex and diverse set of missions and has brought about a heightened need for large-scale coordination.

Furthermore, command and control technology can easily be misused to the detriment of a military organization. For example, modern information handling technologies excel in performing clerical tasks, including counting, collating, and comparing data. Their capacity to do this is so much greater than that of humans that it constitutes a difference in concept, not just degree. However, we can point to numerous examples where we have failed to acknowledge this: where we have applied new technology to old procedures and processes. Further, we often linearly intersperse machine and human functions, whereby the vastly superior clerical capabilities of the machine have actually increased the pace and load of clerical functions still expected of humans.

Avalanches of data are generated by the technology in quantities and forms that the human cannot absorb: seas of computer printouts and congested and complex CRT displays. Ironically, military officers are then forced to work harder at clerical tasks, with even less time to perform the tasks their commander most values: the exercise of judgment.

The research represented in this volume begins with the recognition that our advantage in command and control technology does not automatically translate into advantages on the battlefield. It further represents an effort on the part of key senior officials in the defense community to think of command and control flexibly, iteratively, and holistically and to seek ways to adjust organizations, procedures, and doctrines within our armed forces to better exploit the opportunities that our commanding lead in technology presents.

Stuart E. Johnson
National Defense University

Alexander H. Levis
Massachusetts Institute of Technology

May 1989

1

C³, An Operational Perspective

VAdm. Jerry O. Tuttle

Good morning ladies and gentlemen, leading scientists in command and control. I am delighted and humbled to address such an august group. All of you are the reason that I am in my current position as the Director, Command, Control, and Communications Directorate, in the Joint Chiefs of Staff and why I am here today. Because of your assistance during my forty months as a battle group commander, some people mistakenly thought that I know a great deal about C³ systems. I owe you and your counterparts an enormous debt of gratitude.

Early in my tour as a carrier battle group commander, I recognized the tremendous resource of knowledge and technology that those of you in our DOD laboratories possessed. Until then, my command and control universe had been limited to the unnecessarily restricted radius of 200-300 nautical miles (nm) of the customary range of action for the carrier air wing. I immediately sensed that a carrier battle group should control the battlefield under an arbitrary 1,000–nm envelope. Nevertheless, I was sagacious enough not to specify any distances lest they be considered limits and chiseled in stone. I did not want to accept any limits. From a commander's viewpoint, I had created a major C³ challenge. With the then-existing systems, I might just as well have been in a 55-gallon drum.

Fortunately, and solely by chance, I turned to you in the laboratories to satisfy my requirements, because I realized that the system was too slow to meet my needs during my tour. You responded magnificently. Through frustration and necessity, I discovered the tenets so ably articulated in the National Defense University's and AFCEA's *Science of Command and Control: Coping with Uncertainty* (Stuart Johnson, and Alexander Levis, editors). I recommend it to you for reading. I read only the foreword and introduc-tion, and purposely did not read any further. It was clear that the editors and authors had mas-terfully captured the essence of command, con-trol, and communications. Reading further would have conditioned me to restate uninten-tionally what you already have available. Besides, I have some other insights that I want to share with you.

Frankly, the challenges and responsibilities for resolving those existing C³ shortfalls reside primarily with me and are not due to a lack of technology. Clearly, there are unconquered and undiscovered spheres of research and technolo-gy that remain and existing technology that needs to be applied or matured. The most fruitful areas reside in improving the *system*: in *management*, in changing *cultures*, and, yes, in *leadership*.

I am totally convinced that experts like you can engineer, design, and package C³ systems to satisfy our every recognized C³ requirement, if the users can properly articulate what is needed. That is the gulf we must bridge. As your director for C³ systems, I hope to con-tribute to and exploit the vast capabilities that already exist. I am confident about our prospects. This symposium, as well as recent publications on command and control help to inspire this optimism.

My staff had prepared a "Strawman" for this address. Frankly, it missed the mark because it was prepared by those who were products of the existing system. The laws of physics have not changed since creation. Our creator kindly provided us with a spectrum between dc and light, and you ladies and gentlemen are masters of exploiting this spectrum. For example, fiber optics are an exciting example of a capability that still begs to be fully exploited.

Let me tell you where I have channeled my energies during this past year for your critical review and conclude with an ongoing veritable C³ success story that needs to be told and one in which you should feel pride, because you have made it possible. Permit me to proceed through the salient topics in more or less chronological order.

When I was informed of my current assignment, I thought that I should learn about the World Wide Military Command and Control (WWMCCS) ADP System. To me, the acronym was just letters in the alphabet. While at CINCLANTFLT, I arranged to be given a "Dick and Jane show and tell." An Army Major struggled through a hands-on demonstration, but frequently deferred to a lady who worked on the staff in logistics. She politely provided many of my answers, but I came away with the perception that WWMCCS ADP was murderous to the user, a white elephant, correctly designed but antiquated, and being used by the wrong people. More accurately, the people who should have been using the system were not. The WWMCCS ADP system was not being used to command and control anything.

In order to become familiar with the system, I had a WWMCCS inter-computer network (WIN) terminal installed in my office prior to my arrival. My discovery was astounding, shocking, unbelievable, and depressing. The fastest, most secure and pervasive C³ system was not being utilized by operators. (They did not know how.) No terminals had been installed in our numbered fleet flagships. There were no "yellow pages" (that is, a list of who was on the system) and no catalog of what programs or files were available in the system. Most amazing of all, the Chairman of the Joint Chiefs of Staff could not send his CINCs messages due to a lack of procedures. I discovered that more data were transferred through the ether every night by the Joint Deployment Agency, those concerned with logistics, etc., than went over the Autodin system. Believing in the aphorism that it is better to light one candle than to curse the darkness, I initiated an awareness program, recommended that the Operations Directorate on the Joint and Unified/Specified Commands use the existing system, and openly marketed its capability.

Today, the Chairman can send messages immediately to any or all of his CINCs through a teleconferencing net; OPLANS, OPORDS, warning orders, alert orders, and execute orders are routinely sent to the CINCs and Joint Task Force commanders; all numbered fleet flagships now have WIN terminals; operational security has improved greatly; and the CINCs are now unanimously in favor of the follow-on WWMCCS Information System, or WIS. Today, the commanders of the Central (CENTCOM), Atlantic (LANTCOM), Military Airlift (MAC) and Southern (SOUTHCOM) commands use the WIN system daily for command and control of real world contingencies. This is quite an accomplishment when one considers that some of their operators could not log onto the system as recently as December 1987.

WIS had experienced a turbulent funding history and could not have survived without the successes demonstrated on the WIN network. Despite the rocky road of WIS, we should field a local area network at five test sites early next year; an automatic message-handling system; and the first increment of the Joint Operational, Planning, Execution System (JOPES) or a link-up of the Joint Deployment System and the Deliberate Planning System.

That's the good news! The bad news is that it should not have taken that long, or cost as much as it did, to design and field a local area net and an automatic message-handling system. The blame certainly does not rest with you in the laboratories. However, it does lie with those who failed to state the requirements adequately, failed to include the users from the beginning, imposed incredibly stringent test and evaluation requirements, established an unwieldy organization whereby no one could be held accountable, and then tolerated poor management until recently.

Where are we headed? Hopefully, the release of GCOS-8 will make the WIN far more capable, faster, and user-friendly, and will greatly expand the set of users. WIS will come to fruition and with the assistance of a system that I have put together during the past six years, will define the execution phase of JOPES better.

Being from outside the C³ community when I reported a year ago, I wanted to remain open–minded in order to learn where the problems were while not revealing my ignorance and thereby losing credibility. I became actively involved with the C³ professional groups such as AFCEA, IEEE, AIAA, and EIA. They provided me with valuable feedback. I learned that we were not making our requirements known to industry. We were not making available our stated joint required operational capabilities (MROCs) until we went out for requests for proposals. This seemed myopic to me, particularly in view of the fact that as joint operations become far more fashionable, unquestionably there would be more joint ROCs/MROCs. I wanted industry to be working on what we need, not what they thought we need. We immediately made our ROCs/MROCs available to industry so that their technological bases could be applied to our requirements sooner. Industry, I hasten to add, has been alert to making the necessary changes in how it conducts business since the enactment of the 1986 reorganization act.

By reviewing our ROCs/MROCs for release, we discovered that some had been overtaken by events; for instance, the Rapid Deployment Joint Task Force no longer existed. Most were poorly written. It was taking up to three years to validate requirements. No review was being conducted to measure the effectiveness of the services to satisfy the CINCs' requirements, etc. I found that the directive that governed the ROC validation process was of 1982 vintage and seriously flawed. We have rewritten it to speed up the process greatly, provide the CINCs with a quarterly situation report, and participate actively in all phases of satisfying the requirement from validation to full operational capability.

For openers, I found that we mistook feasibility, practicality, affordability, and priorities for *requirements*. I can count to four. If one of those CINCs says he has a requirement, then this three-star (and certainly not a committee) is not going to tell him he does not have a requirement. If General Piotrowski, CINC-SPACE, says that he needs to talk with Voyager on the other side of Pluto in less than four hours, then he has a requirement. I then come to you to bail me out with a transmission medium that is faster than light.

The entire C^3 operational requirements process has made major progress, but it is still unsatisfactory. Nevertheless, we will prevail. There will be an ever-increasing number of joint requirements as we conduct far more joint exercises. This is good because it will nurture interoperability. Since Grenada, we have come light years in our ability to interoperate. However, it is not enough that the Air Force,

Marines, Army, and Navy are interoperable. We must be able to operate together with our allies while using compatible C^3 systems. This encompasses language, procedures, technologically compatible C^3 systems, doctrine, etc.

We are actively pursuing all of these areas and, frankly, I am pleased with the progress. I do not believe that we are mistaking activity for achievement.

Interoperability must start from standards. We, in J-6, simply were not taking an active enough role in establishing standards and in ensuring that C^3 system developers were rigidly adhering to these standards. I do not want to reveal that embarrassing state of affairs, but we are very much involved in creating standards and ensuring compliance. Some of the more salient areas that we are pursuing are fiber optics, high frequency anti-jam wave form, over-the-air crypto rekeying (more on that later), etc.

Perhaps the most significant accomplishment regarding standards has been the utilization of an existing standard — the message text format (MTF) standard. When, by fiat, JINTACCS MTF became the standard message format for the 132 selected messages on September 3, 1986, for various reasons (primarily ignorance) it was considered that they should be utilized for exercises or contingency operations only. Admittedly, many of the message formats were for shore bombardment, naval gunfire support, etc. Nevertheless, it is foolish to introduce an unfamiliar procedure or system when embarking on contingency operations. Clearly, we had to find a way to familiarize everyone with MTFs by using them routinely.

To provide leadership, I directed that all messages drafted by my directorate be in MTF. The response by the CINCs has been phenomenal. As recently as this past December, for all practical purposes, the Military Airlift Command did not use MTF. On their recently concluded exercise, 98 percent of all messages originated by MAC, and 82 percent of the messages received from their components, were in standard MTF.

What we have discovered by daily use of MTF is that our automatic message distribution systems require reprogramming. This discovery led to the realization that we had three different automatic message distribution codes —

one for Air Force, another for Navy, and yet another for NATO — all of which were inefficient computer programs. As I speak, we are working on a standard, efficient automatic message-routing code for us and our allies.

As an operator, one of my objections to JINTACCS MTF was to an air gap and message format change at the Joint Task Force level. Because the Joint Task Force is the last place that you want a discontinuity, we have now corrected this discrepancy, and the system is transparent from the unit level to the national command authorities. The message I want to deliver with the MTF story is that it is another example of a capability that existed, but was being seriously restricted by ignorance and a lack of procedures. It also provides a convenient lead-in to my next topic.

No system is any faster than its control mechanism. Regrettably, our C³ system is far too dependent on recorded messages. This is unacceptable, certainly for air defense and most other dynamic scenarios. Even those who graduated magna cum laude from Evelyn Wood's speed reading course cannot envision the situation from a paper-based system. By achieving universal use of the machine-readable MTF, at least we will be able to correlate MTF reports with other faster sensor-derived data in order to reduce the uncertainty addressed in the book *Science of Command and Control: Coping with Uncertainty.* It is this uncertainty that I want to ameliorate, having been in command of something for the greater part of my Navy career.

I am confident that, with minimum resources, I can cause numerous existing databases to be fused in order to form a common red and blue database that can be used at all echelons of command — from the national command authorities to the individual missile shooter on scene. The database at each tier in the echelon can be tailored locally by granularity or geography to suit the commanders needs. This tailored composite red/blue picture must be projected in such a manner that it can be conveniently, constantly, and critically analyzed and challenged.

The creation and management of these databases can be made far more automated, and a more equitable division of labor in intelligence production at all echelons can be achieved while reducing redundancies.

Economies can be made in personnel resources, but, more importantly, all concerned will be operating from a common database.

I have initiated this effort in the National Military Command Center (NMCC), which currently has 115 individual systems (a number clearly unmanageable), no composite picture, and no way to project it automatically, if there were one. That is not totally correct, inasmuch as we did throw together a system, literally overnight, in the Operations Deputies Conference Room for the U.S./Iranian encounter in the Persian Gulf. As a result of our being able to project on a large-screen display near real-time locational data for units in the north Arabian Sea and Persian Gulf, Secretary of Defense Carlucci and Chairman, JCS, Admiral Crowe could make far better and faster decisions. As a matter of fact, they could sit back and observe General Crist in Florida orchestrate the entire affair. Of course, we had provided General Crist with a similar capability. With the on-scene commander, RAdm. Tony Less, the CINC (Gen. Crist in Tampa, Florida), and the Secretary and Chairman all having the same picture and same databases, the requirement to communicate diminished markedly. By having the red and blue forces depicted in one composite picture, the relative urgency for decision making could be readily determined and priorities set more intelligently.

How have we arrived at this juncture? Certainly not for the lack of technology. It has been accomplished by knocking down barriers and changing a culture. It started from a Joint Operational Tactical System (JOTS) that had evolved from a desk-top computer, tactical decision aid, and planning tool to a management information system and thence to a C³ system — first for a carrier battle group commander, then for joint task force commander, then for a numbered fleet commander, and finally for a fleet commander. Now we are developing it for the world duty officer.

At this juncture, I want to point out an illness in our systems approach that must be ameliorated. It is natural and easy to see why it happens, and it is the reason why we had 115 separate C³ systems in NMCC. Understandably, any system's requirement is articulated from the perspective of the echelon of command for whom the requirement exists. There is little or no consideration of where the system that satisfies these requirements fits into the whole. The situation is further exacerbated by the executive agent and program manager for the system, because they want and need to have limits on their project.

There is a project rapidly maturing today that feeds from multiple external sensors, but it is a closed system, in order to satisfy a single-attack option when its unique sensor could be used to great advantage by many other weapons system. I must arrange to have this system interoperate digitally with many other systems. The worst examples of this can be found in the intelligence system. I call them "stovepipes." This is not an indictment of the intelligence community by any stretch of the imagination, but, rather, a recognition of the fact that the titular head for all of their systems is a J-2, or intelligence officer—not the officer in charge who has the ultimate responsibility in his Area of Responsibility (AOR).

Currently, intelligence consumes approximately 65 percent of our communications capacity and is expected to increase to 85 percent during a conflict. Clearly, we need to go on a communications diet and analyze data as opposed to handling data. I am working with DIA and NSA in order to achieve a more comprehensive and composite picture for all echelons of command while burning up less of the airways.

I was demonstrating the Joint Operations Tactical System to a JCS operations officer one afternoon. He said: "That's great! Have it in place for tomorrow's first reflagged tanker escort mission." We worked all night getting the connectivity for the JOTS picture back to the Pentagon. We took the officer in tactical command (OTCIXS) JOTS broadcast off the Indian Ocean satellite and had my friends in Naples rebroadcast over the Atlantic satellite to CINCLANTFLT in Norfolk and over landline to me in the crisis action room. Wanting to bet on aces, straights, and sure things, I did a similar "M" hop in the Pacific through Guam and Naval Ocean Systems Command, San Diego.

Now the connectivity is not impressive and, in fact, is routine. But what occurred that morning at 0630, when the Secretary and Chairman watched near real-time the Operation Ernest Will ships steam through the Straits of Hormuz and out of the silkworm envelope, changed a culture. Afterward, the Secretary

picked up the red phone and said: "Chris, super job." He departed saying that this was the first time that we had beat CNN. What General Crist could not figure out was why the Secretary and Chairman had been so quiet and had not asked any questions.

I rapidly became the most popular three-star in DOD. Every four–star whose area of responsibility I had used to pipe a picture of General Crist's AOR to Washington, called me and "reviewed my heritage." General Crist had changed it all, for he intrepidly said that he did not care if his superiors had any, and all, information as long as he had the same information at the same time. We provided him with the same capability that we possessed, and we have grown together ever since.

The other morning, around 0530, it was gratifying to be able to sit there and have the complete picture. However, there is another major issue here. All depicted data were being challenged constantly and by every echelon of command. If the same information had been in message format or in a database, the degree of uncertainty would justifiably have been greater.

I am so committed to developing this system which will provide worldwide tactical C^3 to the NMCC and all the CINCs for contingencies covering the spectrum from low-intensity conflict to theater nuclear war, that I have established an entire division on my staff to work on the problem.

I want this tactical C^3 system to be used continuously, to be reliable and fairly robust, and to become ever-increasingly less vulnerable. It is my "Use, Learn, Develop" philosophy.

To utilize these and other existing C^3 systems better, we have just concluded an over-the-air rekeying for crypto devices worldwide exercise. We demonstrated that we can send a crypto key list over existing communications channels covered by KG-84s and Vinson crypto devices—a capability that has existed for a long time, but had never been utilized. I cannot even imagine the amount of money, trees, and careers that will be saved by this procedure.

As I approach the end of this presentation, you will note that I have not asked you for one thing. The reason is that until I can fully exploit the capabilities that you have already provided, I do not intend to ask for more. Just continue your superior work. By the way, if you solve the multi-level security problem, I would be very interested. We are devoting tremendous resources to this issue and may be demanding of our hardware and software engineers unrealistically lofty requirements and an electronic system to compensate for a lack of discipline in our security system.

We need to work together on two things:

1) Embedded crypto in our C^3 systems and

2) Global Positioning System and weapons systems integration.

Now I would like to conclude with a success story that is largely unrecognized and one for which you are largely responsible. Let us mentally take a trip halfway around the world near the cradle of civilization. There we see French, British, Belgian, Dutch, Saudi, and Italian ships steaming in hostile waters with numerous oil tankers, ships carrying military cargo, Iranian combatants, and (yes!) a sizable U.S. naval presence. All of these link-capable ships are in the U.S.'s and/or Saudi AWAC's link and the E-2C's link, when present. The U.S. ships can communicate on secure channels with the allied ships in the theater, with aircraft for deconfliction, with the reflagged tankers, ashore with all of the Gulf Coast states' Air Forces and Navy headquarters, and with our embassies. Now consider that the CINC who has the ultimate responsibility for this theater resides on the other side of the world, 7,000 miles away. His Navy component commander is in Hawaii; his joint task force commander embarked in his flagship in the north Arabian Sea; and his surface commander is in the Gulf. This is testimony of how a C^3 system can be engineered to support the war fighter. This scenario is even more noteworthy when you realize that this was a virgin C^3 area a year ago and that some who are now communicating over secure circuits would not have exchanged greetings before, if they had met on the street.

Thank you for permitting me to participate in your symposium. I hope that it, like the county fair, gets bigger and better every year. I leave you with this thought: Turbulent progression is preferable to tranquil stagnation.

---

* Editors' Note: This is a slightly edited version of the keynote address given by VAdm. Tuttle at the opening session of the 1988 Symposium on C^2 Research, Naval Postgraduate School, Monterey, California, June 1988.

2

A Discipline for Command and Control

Harold W. Sorenson

Abstract

Is there a discipline of command and control? The military has command and contro l (C²) systems and continues to talk about the need for greater capabilities of these systems. It is not apparent that they understand how to achieve these capabilities beyond the context of wiring more powerful computers together through communication nets with larger and larger bandwidths. There seems to be a widespread feeling among the users that their systems provide less capability than is needed and, based on their perception of the computer revolution, less than is possible. On the other hand, it is apparent from the content of the talks and papers presented for these conferences that there is an emerging discipline. The JDL Conference in Monterey in June 1988 was organized into four major categories that provide a top-level outline of a C² discipline. It seems that the basic question needs to be restated, possibly as: Is the emerging C² discipline addressing the fundamental issues that will lead to a evolutionary and timely improvement in fielded command and contro l systems that will significantly enhance military effectiveness? This paper is intended to provide a personal perspective on this question and to propose some partial answers.

A General View

While a member of the faculty, specializing in Systems Theory, in the Department of Applied Mechanics and Engineering Sciences at the University of California, San Diego, for twenty years, my interest in C² until the past three years was primarily theoretical. After spending 1985 through 1988 as Chief Scientist of the Air Force, my perspective has changed substantially. The opportunity to learn AF operational capabilities and concerns has forced me to reevaluate the role of the R&D community in general and certainly to reconsider the ways in which C² problems need to be addressed. The following represents a desire to explain and to communicate some of the views and conclusions that I have reached. The subject of command and control is vast; only a portion of the problem can be addressed in this paper. For example, only a limited portion of the AF programs in C² are used to illustrate some of the points that will be made.

A brief description of the position of the AF Chief Scientist will provide some basis for the perspective from which my remarks are taken. The Chief Scientist is on the Air Staff and reports directly to the Chief of Staff. The position was established in 1950 and is intended to be filled on a temporary basis by a scientist or engineer, not otherwise employed by the Air Force, and drawn from the research community. I was the 24th in a very distinguished lineage.

Upon taking the job, I was told that the Chief Scientist's mission is to provide "objective technical advice to the Chief, to the Air Staff, and to the Air Force." There is no other charge in the job description. The position by design has no administrative or managerial responsibilities, so the Chief Scientist is free to pursue topics and interests, primarily of his choosing, to whatever depth seems appropriate. In three years I received very few direct taskings although the ones that I did receive were significant. The protocol rank of the office ensured the cooperation of all branches of the Air Force and of the industry that supports it.

Although I had worked with the Air Force and other agencies of the Department of Defense throughout my professional career, the breadth of the charter, the level of involvement, and the size and scope of the AF R&D program seemed overwhelming at the start. Being unfamiliar to a major extent with the operational AF, it became obvious that it was important to travel to operational bases to learn about the user requirements firsthand. Then, the capabilities that the R&D community had provided to the operational Air Force would become more tangible, and it would serve as a basis for identifying technology areas deserving greater attention. Since the stated posture of the Defense Department is to use technology to overcome the numerical advantages of potential enemies, an appreciation of the ways technology is supporting this posture is important.

It did not take many visits to operational bases and commands to arrive at a conclusion that has only been strengthened with time. There appears to be a dismaying gap between the potential that technology offers and the capabilities that are fielded. But with time, this impression was tempered by the realization that advanced technology has contributed to an impressive military capability in many areas. Certainly, the F-15 and F-16 fighter planes exhibit remarkable capabilities and their performance must be regarded as triumphs of our technology. The intercontinental ballistic missile fleet and their impact accuracies are nothing short of amazing to someone (i.e., the author) who began his career working on the Atlas missile. Many satellite systems perform electronic miracles in a routine and continuing fashion. Regardless of the claims in the press, the B-1B bomber represents a remarkable advance beyond the capabilities of the B-52. One can continue to add to the list of commendable achievements of technology that do provide this nation with an impressive military defense.

Nonetheless, there are areas where technology does not appear to support users to the extent that is possible. With considerable support from many people in the Air Force, I identified several "chronic problem areas" in which there has been a steady R&D investment for many years with a less than adequate return in terms of military capability. Without trying to be all-inclusive, these chronic problem areas include Electronic Combat (EC), Identification-Friend, Foe, or Neutral (IFFN), Autonomous Stand-off Weapons (ASOW),

Remotely Piloted Vehicles (RPVs), and Command and Control/Battle Management (C^2/BM).

There are difficult technical challenges (e.g., the automatic target recognizer for ASOW) for these problem areas but the biggest challenge, and a common theme in each, is the difficulty, even the inability, to assess the effectiveness of the system and to measure its contribution to war fighting capability. Unlike the speed and maneuverability of a fighter aircraft, it is difficult to imagine or understand the effect of these systems on the outcome of a battle. For example, the value of an EW jammer against a single radar can be determined, generally with considerable difficulty because of uncertainties about the features of the radar, but it is extraordinarily difficult to measure the effect of several jammers operating against many radar systems which are part of an integrated air defense system. As another example, the ability of an ASOW to attack a specific target may be assessed but the greater questions concerning its planning, use, and effectiveness against a realistic environment is poorly understood. Since the cost of an autonomous weapon is orders of magnitude more than a "dumb bomb," a realistic measure of the weapon's effectiveness is imperative.

To repeat, it is my conclusion that there is a common theme among these chronic problem areas and this theme is the fundamental contributor to the slowness with which progress is being made toward their solution. The link (or cause-effect relationship) between military utility for realistic situations and the capability provided/needed by the system/device has not been established. Furthermore, the problem with realistic assessments of the potential contributions of such things as ASOWs, RPVs, EC systems, and IFFN systems comes from an inability to understand and describe the manner in which the command, control, communication, and intelligence systems can or do influence the management of a battle or a war. While one can dismiss this as a restatement and re-recognition of the "fog of war," this seems an invalid response. Unless the fog can be reduced through rational evaluations which consider the unknowable aspects of the war fighting environment, these chronic problems will continue to frustrate and perplex leaders, planners, and implementers as systems are produced which do not meet the needs of their users. And at the foundation of this process of increased understanding is the need to understand the role and effectiveness of the command and control system on the management of a battle.

A Cursory Review of AF Tactical Command and Control

Why does the military need a command and control system? Certainly, the process of command and control is needed to plan and execute combat operations necessary to deter or fight a war — and win. The resultant collection of procedures to perform this command and control process may be thought of as the command and control system. Command and control is accomplished in the Air Force today and command and control systems have been evolved to support this process. It is worthwhile to review the process by looking at the major components of the process: recording current status, knowing about opposing forces (i.e., intelligence), developing plans, disseminating the plans, executing those plans, and then analyzing the effectiveness of those plans. Modifying the plans may be thought of as a next iteration of the aforementioned process.

One of the most fundamental aspects of command and control is knowing the current status of friendly forces. Currently, this status information is recorded using grease boards and pencils. Updates from the field are usually transmitted via voice communications or message traffic; any changes are recorded on the grease boards. A request for information from other headquarters results in the appropriate information being read from the status board and communicated, usually, via radiotelephone where it is copied manually onto their current status board.

The successful conduct of military operations requires that as much as possible be known about the opposing forces' capabilities and intentions. This has classically been the domain of the intelligence community . The data comes from a variety of sources: aerial reconnaissance, satellite imagery and sensors, interception of electronic signals (SIGINT), ground-based sensors, reports from agents and friendly patrols (HUMINT), etc. These data are sent to the theater and/or national level where they are interpreted. It is this interpretation that is currently referred to as the *fusion* problem. Tactical and operational intelligence estimates are then formulated and transmitted back to the appropriate units. Each unit at the next lower level receives the information necessary for its area of responsibility (with some overlap with its neighbor's). These units then perform the same intelligence breakout for their respective subordinate units and so on. There are important questions associated with the distribution of this information up and down the chain. Who should get what and when do they have to have it? At present one gets the impression that everyone gets everything whenever it becomes available.

Based on the mission, the current status of friendly forces, and the current intelligence, the commander and his staff develop the operational plan (OPLAN). In times of peace, various scenarios are developed and operational planning for each is accomplished. Should a conflict arise according to one (or more) of the envisioned scenarios, the appropriate OPLAN is executed. The operational order (OPORD) directing execution of an OPLAN is typically of the form "Execute OPLAN xxx as OPORD xxx effective...". Even if the actual situation does not match any of the OPLANs, there may be enough similarity to one or more of these plans that the actual OPORD only requires these pre-planned OPLANs to be appropriately modified and merged. During actual combat, the development of OPLANs and execution of OPORDs are much more closely related, but the process is essentially the same. The mission is analyzed and subordinate units are tasked. Resources are allocated to subordinate units so they can develop their plans. The commander determines his concept of operation and estimate, various courses of action are developed and analyzed, and the decision is made. Most of this phase of the process is implemented using the classical "stubby pencil work."

The dissemination of OPLANs/OPORDs is usually accomplished by radio, telephone, message traffic, or in person. During actual combat, the time for a commander to plan an operation may only be the time it takes him to travel from his superior's headquarters (where he received his mission tasking) to his own headquarters. Given the mobility provided by modern battlefield assets, this travel time may be short indeed.

Finally, each echelon monitors the execution of the operation by its subordinates. Either the accomplishment of the mission and achievement of the objective or heavy enemy opposition, which slows the friendly forces, will cause the next iteration of the aforementioned command and control process.

It is obvious that the highly labor-intensive command and control process can be assisted by judicious use of computers. First, office automation techniques may be applied to ease the paperwork and associated information flows of this process. One can readily conceive of the application of database management systems and spreadsheets to replace the grease boards and pencils. These would streamline the retrieval of necessary data and information, reduce redundancy, and improve the integrity of the data — all possible today with existing off-the-shelf technology! The development and application of distributed database systems would aid the real-time availability of data at several echelons. This requires the solution of some significant technological problems in the areas of operating systems, distributed database management systems, and computer networking.

The intelligence/data fusion problem needs to be addressed. Automatic analyses of incoming sensor data could be transmitted in real-time to update the status of opposing forces and/or friendly assets. While sometimes the commander will wish to see the detailed data for himself, most often he will need only information that provides a broad view of the situation. Consequently, it is important to understand the intelligence information requirements at each level of the command structure. There is no need to swamp everyone with information that may not be useful for their purpose and mission. In addition, this system needs to be capable of operating in a man-in-the-loop mode, thus necessitating an easy human-computer interface and its associated user information management system.

At still another level of application of computing technology, artificial intelligence, expert systems, and simulation techniques can aid the command and control process. Expert systems could be designed and implemented to aid the user in using the system, to simulate and war game the various courses of action developed, and to aid the user in generating the OPLANs/OPORDs. These techniques would allow the user to query the system to determine how a particular system-generated plan was derived.

It does not take very long to identify tools that could expedite the manner in which the command and control process is accomplished at this time. Many improvements are being introduced through the utilization of off-the-shelf software packages that make it faster and easier to record information and to perform simple trade-off analyses. In the next section, efforts in this direction that please and excite the user community are addressed. Few of these efforts have been initiated by the development community, although they have become increasingly involved as the problems of a piece-meal approach become more apparent.

Some Tactical Battle Management/ Command and Control Efforts

To provide a background for some of the assertions made in the next sections, I will briefly review some current tactical battle management/command and control efforts. These examples define major development thrusts and illustrate some general points. First, let me remind you that there are major differences between strategic and tactical command and control.

The strategic problem allows the luxury of deliberate planning. Plans can be developed for several alternatives; the plans can be "war gamed"; and options can be selected or rejected as appropriate. The execution phase of this system must be implemented only once — and we earnestly hope never! On the other hand, the tactical problem is crisis-oriented, extremely time-sensitive, and spans the entire spectrum of conflict.

In 1986 the Air Force Studies Board conducted a Summer Study on "Tactical Battle Management." The objectives of the study were to evaluate the potential for providing information management technology to tactical battle management functions. The findings of the summer study are summarized with the following statements.

• Tactical battle management capabilities

can be significantly improved using *current* technology.

• A *systems engineering* approach, involving both users and developers, is needed.

• *Evolutionary* improvement, with *rapid prototyping/fielding*, is key to improving tactical battle management with advancing technology.

• More *advanced development* support is needed to bridge the gap from laboratories to the field.

The Air Force Study Board recommended the use of evolutionary acquisition for selected programs and the centralized management of tactical battle management initiatives. The study also recommended the development of a user/developer test bed. The subsequent formation of the Tactical Battle Management General Officer Steering Group (TBM GOSG) is indicative of the high-level support this initiative received. The emphasis of this steering group was on air surveillance management and control, ground surveillance management and control, and force planning.

There are a number of initiatives underway at the present time that support the conclusions and recommendations of the study. Some of these efforts predate the study and reflect to some extent the frustration of the operational commands with the acquisition process and the R&D community. Systems are now being developed using on-site prototypes built with extensive user involvement. The development and use of on-site prototypes is justified by having the goal of reducing the time from the laboratory to the field. These are being used in on-going efforts to provide wing and squadron automation, force planning automation, sensor networking, and tactical C³I test bed interoperability. As examples, the flying unit automation efforts center on development of a mission support system and a wing command and control system (WCCS). Force planning automation efforts include development of a Contingency Tactical Air Control System , a tactical expert mission planner, a C³ Countermeasures Decision Aid, and multi-media communications.

There is strong evidence that with on-site efforts users can gain, via rapid prototyping, hands-on experience sooner and can field selected capabilities faster than through standard acquisition procedures. It is also apparent that operators are willing to live with a less-than-perfect system initially, if it is *their* system (developed by them and/or tailored to their needs). It is also clear that the operational developers are not especially good integrators and do not always have a perspective regarding what alternative solutions may be available or what is being developed elsewhere. The R&D community can provide this perspective and the observation emphasizes the need for a closer relationship between developers and users.

The General Officer Steering Group recommended the formation of a Tactical Battle Management Initiative, and it is useful to review a couple of the individual efforts being developed under its aegis. Two major examples are the Advanced Planning System (APS) and the Wing Command and Control System (WCCS) Prototype. The Advanced Planning System is a decision aid being developed by the Rome Air Development Center (RADC). APS will be the first large-scale application of the Tactical Battle Management Initiative for rapid technology transition from the laboratory to operational users via close user-developer cooperation. The test bed is located at the headquarters of the Tactical Air Command at Langley Air Force Base, Virginia. As such, it implements the approach briefed at the July 1, 1987, General Officer Steering Group and as discussed and approved at the two subsequent working group meetings.

Initial tactical battle management surveys revealed capability gaps in the ability of the Tactical Air Forces to do force-level planning at the Tactical Air Control Center (TACC). Specific examples include gaps or weaknesses in accomplishing planning for the use of electronic combat capabilities and in the ability to do weather-based predictions of precision-guided munitions effectiveness. At a fundamentally important level, the generation of Air Tasking Orders (ATOs) continues to be done in a man-power intensive manner (e.g., grease boards and pencils) and there is a clear need to use computer automation to a greater extent and in a more integrated fashion than is currently fielded. Current systems capabilities have been developed by different agencies and use different hardware, operating systems, database management approaches, and display technologies. The problem is further complicated by differences in tactical air force theater command and control systems hardware. There are developmental efforts that address separate parts of the overall ATO generation problem but the need to integrate them is now recognized.

The approach being taken in the Tactical Battle Management Initiative tends to capitalize on existing investments and extract the necessary functional capabilities from each piece. There is the intention to integrate the capabilities from the following existing software systems:

• Force Level Automated Planning System (FLAPS)

• Improved Many-On-Many (IMOM) simulation

• C3 Countermeasures Battle Management Decision Aid

• Weather Tactical Decision Aids

• Tactical Expert Mission Planner (TEMPLAR)

FLAPS was developed by U.S. Air Force Europe (USAFE) for use at the Air Tactical Operations Center (ATOC) level. It matches weapons and aircraft to targets for Air Interdiction and Offensive Close Air portions of the ATO. It determines the most survivable penetration routes through known enemy defenses.

The IMOM simulation was developed by the Air Force Electronic Warfare Center as an in-house effort. It is in use as a stand-alone system at a number of locations. It analyzes the effectiveness of stand-off (EF-111) and fighter aircraft self-protection radar jamming systems. Its graphics displays are readily understandable, especially when superimposed on proposed penetration routes.

The C3 Countermeasures Battle Management Decision Aid, developed by RADC, uses expert systems techniques for disruptive versus destructive option selection. It provides integrated analysis of radar jamming and communications jamming effectiveness and assesses the effects of friendly jamming on our own C3.

The Weather Tactical Decision Aids System was developed by the Air Weather Service and predicts precision-guided munitions (TV, IR, laser) target acquisition and lock-on performance. Based on forecasts for various weather parameters, including aerosols and absolute humidity, it can help force-level planners determine suitability of including precision-guided

munitions tasking in the ATO.

The Tactical Expert Mission Planner, developed by RADC, uses expert systems techniques for planning an entire ATO. It provides constraint checks and package deconfliction.

The WCCS prototype started as a USAFE initiative to provide secure, distributed command and control processing within USAFE operational flying wings. The initial site, Spangdahlem Air Base, Germany, proved the feasibility of using a secure wing command and control system. However, because this installation does not include distributed processing, a single point of failure exists for the entire system. The WCCS prototype at RAF Lakenheath, England, will rectify this situation, if successful, by incorporating distributed data processing features to enhance wartime survivability. This effort also will evaluate the software developed during the prototype for use as the basis for TAF common software. This prototyping effort will incorporate the functionality of three operational systems. These three functions automate squadron operations, including squadron scheduling; support daily flying operations; and automate air base survivability and critical resources monitoring. As much developmental work as possible will be done in SQL to facilitate conversion of prototype efforts to an operational system. The common core capabilities of the WCCS will include the ATO breakout and unit scheduling functions and will display facilities status, personnel status, resources status, weather, and intelligence information.

The Air Force tactical command and control problem is only part of an even bigger problem. Clearly, the Army and Navy have their own problems and have initiatives to solve them. Consider briefly DOD initiatives that address other aspects of the overall military command and control problem. Perhaps best known, the WIS (WWMCCS Information System) effort is a multi-faceted initiative to upgrade our top-level command, control, and communications capabilities. Another effort is Level 6 theater command and control systems efforts of the Defense Communications Agency (DCA). The way in which we will have to fight (if the time does ever come) is with combined or joint operations. The most important command and control issue in this environment is the problem of interoperability.

The Joint Interoperability Tactical Command and Control System (JINTACCS) being developed by the Joint Tactical Command, Control, and Communications Agency (JTC³A) is working very diligently to address this problem. A monumental effort to define standard interfaces is on-going.

A Discipline for C² Systems Development

The preceding review of the AF Tactical Battle Management Initiative is intended to provide support for the more general discussion of this section. The examples that have been discussed were chosen to illustrate and, hopefully, make plausible some of the assertions that follow. The conclusions are based on both a broader and deeper review of Air Force activities.

There is no doubt that the Air Force recognizes the importance and, therefore, the need for the greatest possible capability in its command and control systems. There are command and control systems in existence and substantial outlays have and continue to be made to procure greater capabilities. If technology is to be the advantage that the U.S. needs to overcome a numerically superior enemy, then technology in support of C² systems must enable the most effective use of the technology that is implemented in our weapon systems.

Computers and communication systems constitute the technologies most often identified as the foundation stones of C² systems. Given the explosion in both computer and communication technologies, it seems obvious that these technologies should permit the development of much greater capabilities than are currently fielded. In fact, it is so obvious that people in the operational forces are frustrated by the apparent inability of the development community to field systems that even approach the capabilities that they can imagine, given their involvement with personal computers. This frustration has led to many self-help programs that actually solve some of the immediate, day-to-day problems that are faced routinely. An example that is widely quoted is the LOCE (Limited Operational Capability - Europe) system which was introduced by USAFE to provide some automated support for the fusion of intelligence reports and data. This system was resurrected from a development program (i.e., Beta) that was cancelled because it did not meet the development requirements. But its limited capability apparently is greater than existing systems fielded in Europe.

In developing the Tactical Battle Management Initiative, it became apparent that the self-help efforts tend to lack focus and to duplicate effort. Further, they generally do not address, adequately, interface and integration issues that arise at other than local levels. The same criticism can be made of many of the development efforts. There seems to be an emphasis on the details of the solution to a very specific need without adequate consideration of broader considerations. This leads to an implicit design approach that is bottom-up and gives very little attention to top-down considerations required for the design of the "system of systems" that is the reality of a C² system. In other words, a disciplined systems engineering approach must be adopted for the design of C² systems.

Air Force Systems Command states publicly and frequently that the biggest problems it has in fulfilling its fundamental responsibilities of developing and acquiring new systems are schedule delays and cost overruns caused by software difficulties. Software is certainly regarded as the most stressing issue in most C² system developments. As a result, software has been given a great deal of attention by DOD and by the Services. In DOD, this has resulted in the development of the Ada software programming language and the Ada Joint Program Office (AJPO), in the establishment of the Software Engineering Institute (SEI) at Carnegie-Mellon University, and in the funding of the Software Technology for Adaptable, Reusable Systems (STARS) program.

Software development is a specific type of a systems engineering problem. Through the attention given to software problems, a discipline has developed that appears to have considerable utility for addressing C² problems. The ideas of structured analysis, design, and information modeling appear to have direct applicability to C² systems. Problems attendant with the development of software specifications from system requirements suggest the need for similar tools and languages for C² system specification. The use of Petri nets for the analysis and design of C² systems appears to be very exciting and is a tool drawn from computer science. These and many other software tools are useful for modeling large and complicated systems and need further exploitation for C² systems. However, software problems seem far from resolution so the use of lessons learned from this community cannot be regarded as providing the solution that eliminates the need for a major effort into the development of a C² discipline. It may be a case of the lame leading the blind.

A great deal of money has been spent for the command and control capabilities that are currently in use. Our community is seeking a disciplined approach to guide and, hopefully, to speed the fielding of greater capabilities. Thus, we have the obligation to address the evolutionary issue of taking systems that exist and improve them to meet a desired level of capability. It does not seem realistic to expect that the ideal system, even if it could be defined, can ever be implemented in a revolutionary way. There are constraints imposed by the developments of the past and the systems of the present and near future.

The design of a new fighter plane, for example, is now an evolutionary process. The fighter planes developed in the past are described in a design database that is used in any new development process. Tests conducted on a variety of aircraft are used throughout the design process and guide the development of test plans for new aircraft. Compared with C² systems, the design and development of new airplanes is a mature and well-understood discipline. Furthermore, airplanes have to obey Newton's Laws of Motion; that alone impose an immediate discipline on the design that cannot be debated or flouted. C² systems do not have Newton's Laws to guide their development, which implies that the need for experimental research is even greater. Test beds and prototypes become an essential part of any systems engineering process used to guide the development of C² systems. It also suggests that researchers need to be very close to the user community and have a clear understanding of the problems and limitations of existing systems. A database of past experience is required to ground the ponderings of researchers and theoreticians.

Newton's Laws provide models for airplanes and missiles that permit the evaluation of performance to be done in a logical and scientifically sound manner. Consequently, assessments can be made of system performance that can be used as a basis for assessing the utility of an airplane or missile in a war fighting environment. But there are limits to this sort of assessment that return to questions of command, control, communications and intelligence. There do not seem to be models of the war fighting environment, including these important considerations, that permit an acceptable assessment of the military effectiveness of many systems and certainly not of alternative implementations of C³I systems. There can be little debate about the desirability of having models and simulations to assess system utility and force effectiveness. Questions associated with utility and fidelity of war gaming models and simulations have long received attention and seem to be receiving increased attention at the present time. There are grave concerns about the R&D communities' understanding and ability to model and simulate the military environments necessary for meaningful assessments of C² systems.

Related to the question of developing models and simulations is the matter of their validation. Exercises and open air tests provide the closest things to a real war that is available. Information gained from these activities must be incorporated systematically into models. For example, the successful development of ballistic missiles, even with well-defined dynamical equations to describe the motion, has made heavy use of post-flight analysis to validate the models. Ballistic missile flights are expensive and the missiles are not retrievable. Thus, there must be great confidence in digital simulations of the missile flight so that a test only validates the "truth" provided by computer analysis. The development of C² models needs a similar application of the scientific method to build credibility.

Conclusions

The assertions that have been made are summarized briefly to highlight the major features of the perspective that has been presented. Underlying the summary is the conclusion that the research and development community and the users of their products must work closely together to ensure that what is possible is fielded. In a sense the users are the universe from which the R&D community must induce the general laws that govern the behavior of an operational command and control system.

• There is an operational pull. We need the best possible C^2 systems if we are to exploit the technology advantages of the United States.

• Current hardware technology in computers, in terminal graphics, and in communications is more than adequate to provide vastly improved C^2 capabilities in fielded systems.

• There is a great need for a disciplined systems engineering approach to the design and development of C^2 systems. This approach must combine the advantages of top-down and bottom-up design to account for existing and local needs and to produce an integrated system.

• Tools for the support of the systems engineering approach are required. It appears that many of the tools and lessons learned from the related systems discipline of software engineering are directly applicable.

• It does not seem possible at the present time to measure the effectiveness of C^2 systems in realistic military situations. This transmits itself to an inability to state meaningful requirements and to measure the effectiveness of other important classes of systems (e.g., EC, ASOW, IFFN).

• A key to effectiveness assessment is the availability of digital models and simulations that represent with a useful degree of fidelity a two-sided war fighting environment.

• A theory and practice of model development, validation, and utilization is as a fundamental component of the discipline. These models have to serve as the Newton's Laws for C^2 systems.

Acknowledgement

This information was originally prepared as a talk at the 1988 Symposium on Command and Control, held at the Naval Postgraduate School, June 7-9, 1988. I would like to acknowledge the substantial contributions of Murray R. Berkowitz, George Mason University and Major, U.S. Air Force Reserve, in the preparation of the material and its inclusion in this paper.

3

Providing Better Support to the CINCs

Edward C. Brady and Stuart H. Starr

Introduction

Recently the Scientific Advisor to the Supreme Allied Commander Europe (SACEUR) wrote to a member of the Joint Directors of Laboratories (JDL) to comment on the most recent JDL C^2 Research and Technology Program [1]. Based on discussions with SACEUR's staff he concluded: "The general feeling was that the program plan focused on the theoretical aspects and was too detached from real world C^3 combat effectiveness. We need to understand and develop C^3 technology to support operational needs. Thus, this program needs to be a balance of theory and empirical measurement." These observations pose a critical question for the C^2 research community: How can it better support the Unified and Specified (U&S) CINCs?

This paper describes the evolving responsibilities of the U&S commanders. Emphasis is placed on the enhanced role of these commands arising from recent reorganizations in the DOD and the limitations they face in executing these new responsibilities. The position of the C^2 research community is then characterized to include a description of the community's current support to the U&S CINCs and a projection of the role that it could ultimately play. Based upon these assessments, several actions are proposed to make the C^2 research community more responsive to the needs of the U&S CINCs.

CINC Responsibilities and Needs

Figure 1 depicts the peacetime locations of headquarters for the ten U&S commands. The composition of these commands has changed over the last few years with the creation of U.S. Special Operations Command (USSOCOM)

and U. S. Transportation Command (USTRANSCOM); the assimilation of the Military Airlift Command (MAC), Military Sealift Command, and Military Traffic Management Command; and the disestablishment of U. S. Readiness Command.

There are several characteristics of the U&S CINCs that affect the support they require. First, the U&S CINCs can be characterized by the nature of their mission: that is, war fighting or support. For example, USTRANSCOM supports the other CINCs in the execution of their operational responsibilities and this support role strongly affects their C^3 needs. Second, the U&S CINCs can be distinguished by the locations of their headquarters vis-à-vis their theaters of responsibility. For example, U. S. Central Command (USCENTCOM) has its headquarters in Tampa, Florida, while its area of responsibility is Southwest Asia. This has strong implications for its C^3 infrastructure. Third, the level of C^3 system support from the military Services and defense agencies varies significantly. For example, close ties exist between the Strategic Air Command (CINCSAC) and the Air Force and between the Atlantic Command (CINCLANT) and the Navy. In addition, the Defense Communications Agency (DCA) has a European office collocated with U. S. European Command (USEUCOM), and a Pacific office with U. S. Pacific Command (USPACOM). However, there are cases, most notably U. S. Southern Command (USSOUTHCOM), where organic and external support for C^3 systems are extremely limited. Fourth, it should be recognized that USSOCOM is the only command with authority to program, budget, develop, and acquire the materiel, supplies, and services unique to its mission.

Each U&S command must be analyzed indi-

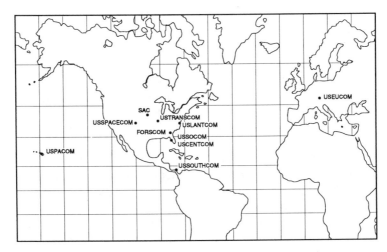

Figure 1. Peacetime locations of U&S CINC headquarters.

vidually to identify the unique assistance required. However, as shall be explained below, there are many common areas of need.

Changes in organization of the DOD have enhanced the roles of the U&S CINCs in four areas: 1) planning, programming, and budgeting; 2) requirements generation; 3) development and acquisition; and 4) training. This section reviews those initiatives, emphasizing the impact on C³ activities and the increased demands they place on the CINCs.

The DOD allocates resources through an integrated Planning, Programming, and Budgeting System (PPBS). The enhanced role of the CINCs in the PPBS is discussed below.

The CINCs must submit a Command, Control, and Communications System Master Plan (C³SMP) as their primary vehicle for identifying C³ deficiencies[2]. The C³SMP documents doctrine, missions, current and target capabilities and systems, an architecture, and an improvement plan to move from the current to the target system in accordance with the approved architecture. In addition, each CINC must contribute to a JCS-sponsored Global C³ Assessment that provides a mission perspective of all systems, at all levels of conflict, and all echelons of command. [3]

The Deputy Secretary of Defense (DepSecDef) enhanced the role of the CINCs in the planning process by requesting that they submit annual integrated priority lists (IPLs) [4]. They were directed to define requirements in broad mission and functional areas and to suggest solutions in terms of required plat-

forms, systems, and items. Although the IPL transcends the domain of C³, it is notable that C³ issues were generally high on each list.

To enhance the voice of the CINC in the allocation of resources, the DepSecDef established a role for them in the review of the Service and Defense Agency Program Objective Memorandums (POMs) [5]. The CINCs meet with the Secretary of Defense and the Defense Resources Board (DRB) to present their views on strategy and to discuss the adequacy of the POMs to meet that strategy. The Services have been directed to develop POM annexes that estimate the costs and to identify shortfalls in satisfying those CINC needs. In addition, CINCs may be represented on POM issue teams.

The Goldwater-Nichols DOD Reorganization Act of 1986 [6] directed that, effective with the FY89 budget, a separate budget proposal be submitted to Congress for activities of each U&S command. The separate funding requests may be for joint exercises, force training, contingencies, and selected operations.

In the past, the CINCs have submitted their requirements in the form of Required Operational Capabilities (ROCs) and there has been widespread dissatisfaction with this process. There was concern that the generation and validation of these ROCs took too long and there were too many and overlapping requirements. CINCs submitted ROCs that reflected solutions rather than statements of operational need.

"Policy and Procedures for Management of

C^3 Systems" [2] was issued to streamline and rationalize the requirements process. The most significant change stems from the modification in the CINCs' role. The CINCs are to provide a non-technical Requirements Submission (RS) that should not exceed four pages. The RS is to document the deficiency, explain how it adversely affects the commander's concept of operations, and prioritize the requirement. In this new process, the validation of this RS is to take no more than ninety days. Technical analyses/cost estimates are to be performed by the Services or Defense Agencies. A ROC that survives this process is considered validated.

A more revolutionary change in the requirements generation process may be in store if the recommendations of the Packard Commission [7] are more fully implemented. Historically, the CINCs have developed requirements from a "bottom-up" perspective, drawing extensively upon their experiences in crises and exercises. The Packard Commission recommended that a complementary "top-down" perspective be taken in which the CINCs would support the CJCS in appraising the threat, developing military objectives, developing national military strategy, identifying five-year force/capabilities within fiscal constraints, developing military options, and preparing net assessments.

In 1980, the CINC Command and Control Initiatives Program (C^2 IP) was initiated to give the CINCs limited discretionary resources to acquire critically needed C^2 systems. This initiative has given the CINCs a small, but highly leveraged, role in systems development and acquisition. As noted by Gen. Rogers, former SACEUR, "since the program began in 1980, at least 200 small but critical projects have been funded" [8]. Under its aegis, several CINCs have developed unique systems to support pressing needs, using off-the-shelf technology. In addition, selected CINCs have instituted test beds to provide early fielding of critical capabilities and to enhance the dialogue between the development and operational communities. One such test bed is the Limited Operational Capability Europe (LOCE), which provides an initial capability for managing sensor systems and fusing their outputs.

USSOCOM has been placed in a unique position to develop and acquire materiel, supplies, and services that are peculiar to its mission. Currently, it is delegating some of this responsibility to Service organizations. For example, one of its high priority programs is the Joint Advanced Special Operations Radio Systems (JASORS), a small, lightweight, secure radio with low probability of intercept features. USSOCOM is the proponent and controlling organization while the Army is Executive Agent and Program Manager. USSOCOM plans to assume a broader role on selected future acquisition activities.

The Goldwater-Nichols DOD Reorganization Act [6] emphasized that the CINCs would have "full operational command" over all assigned forces. However, it defined that term to include all aspects of military operations and joint training; functions not traditionally thought to be included under operational command.

The initiatives cited above have dramatically expanded the responsibilities of the CINCs. There is a perception, however, that the CINC staffs will experience significant problems in responding.

Although the primary focus of this discussion is the J-6 (C^3) element of the staff, the full resources available to the CINC must also be considered; most notably, the J-2 (intelligence), J-3 (operations), and J-5 (plans).

In general, the staffs of the U&S commands focus on the near-term: to the matters of training and readiness. Their most notable shortfalls have been in three interrelated areas: they generally lack in-depth knowledge of key factors that will influence the future (e.g., the evolving threat, capabilities, and limitations of emerging technologies). Also, they generally lack the skills needed to develop credible long-range plans and architectures and they have difficulty communicating with the technical community. Although these shortfalls could be ameliorated by selectively increasing staff size, the CINCs are under considerable pressure from OSD and Congress to make appreciable cuts in staff.

Another major obstacle is a lack of access to key data. This is particularly apparent in the programming and budgeting review processes, where the database is volatile. It is difficult for the CINCs' staffs to formulate or assess issues when they lack timely, detailed programmatic information.

The CINCs need a broad range of tools to evaluate C^3 effectiveness, formulate require-

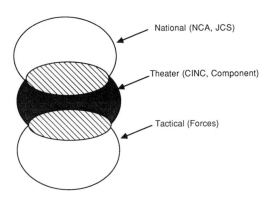

National (NCA, JCS)

Theater (CINC, Component)

Tactical (Forces)

Figure 2. Issues: CINCs area of responsibility.

ments, perform program assessments, and accomplish realistic joint training. Help is specifically needed to formulate requirements. To properly perform this function, the CINC staffs must balance operational needs, available resources, future threat, doctrine, force structure, and evolving technology. Without adequate tools, it is difficult to ensure that all of these factors are properly considered.

Finally, the CINCs face problems in dealing with development and acquisition. As noted above, some commands (notably USCINC-SOUTH) lack adequate linkages to the Service system commands. In addition, nearly all of the CINCs face substantial shortfalls in their ability to integrate new components into their evolving C^3 system and in the support to train for the unique, CINC-developed systems. The CINC role in development and acquisition is further clouded by uncertainty over their domain of responsibility. While they clearly have responsibility for their own organic C^3 resources, there is ambiguity as to their role in systems associated with the national-theater interface, the theater-tactical interface, and tactical forces (See Figure 2).

The CINCs require new skills or outside assistance in these areas: planning, programming, and budgeting; requirements generation; C^2 system development and acquisition; and training.

If the CINCs are to participate effectively in the PPBS process, they must be able to evaluate the contribution of C^3 to mission effective-

ness. This need has been recognized in a recent request from the Scientific Advisor to SACEUR to the head of the Defense Science Board [9]. They also must perform cost analyses of C^3 systems and conduct cost-effectiveness trade-offs. They require organic resources to assist them in developing CINC-unique products and in monitoring compliance with these plans and architectures.

The CINCs need complementary bottom-up and top-down tools to generate C^3 requirements. Currently, many of their requirements arise from anecdotal experiences from exercises and crises. They need to generate credible "lessons learned" from those experiences to justify operational needs. In addition, they require mission-oriented assessment tools to enable them to comply with the spirit of the Packard Commission recommendations for a top-down requirements process.

Nearly every recent "blue ribbon" panel study of C^2 management (e.g., [10] has identified the communications gap between the operational and systems communities as the major obstacle to effective C^2 development and acquisition. To bridge that gap, the CINCs need the ability to conduct an effective dialogue with the systems community. Test bed environments are one method, but that requires experts in computer hardware and software, modelers, experimental designers, and human factors specialists. In addition, the CINC faces the problem of assimilating new C^3 components into his evolving system. This requires skills in managing interoperability, ranging from knowledge of protocols and standards, through changes in procedures and concepts of operation.

New responsibilities in joint training pose unique problems in the area of C^3. It is particularly difficult to realistically emulate the stresses that joint/combined staffs must confront in combat. Test beds and war games have been proposed, but this requires yet another set of unique skills.

Help from the C^2 Research Community

How can the C^2 research community help? As a foundation for this discussion, a taxonomy is introduced to characterize the C^2 research community and to describe the work that the community is currently performing to

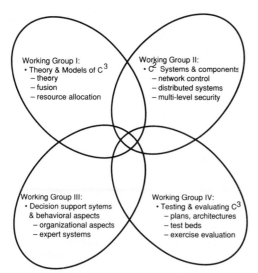

Figure 3. Decomposition of the C² Research Community.

support the CINCs. That is followed by a set of challenges that the C² research community should confront.

The C² research community is comprised of highly non-homogeneous, eclectic participants who have defied prior attempts at classification. However, the structure of the 1988 JDL C² Research Symposium provides a useful taxonomy for discussing the community's interests and describing its activities.

The symposium has subdivided the community into four overlapping domains (See Figure 3). Working Group I is involved in the theory and models. It is striving to establish a theoretical foundation for C³ and is performing applied work in areas such as sensor fusion and resource allocation. Working Group II is focusing on state-of-the-art enhancements to selected C² systems and components. Major subareas of interest include network control, distributed systems, and multi-level security. Working Group III is analyzing behavioral aspects of C³ and developing decision support systems. Their domain of interest subsumes the organizational aspects of C³ and expert systems. Finally, Working Group IV is exploring means of testing and evaluating C³. This includes the development of C³ plans and architectures and the development and application of the full spectrum of evaluation techniques (e.g., analyses, test beds, exercises).

These four segments of the C² research community are currently involved, by varying degree, in supporting the CINCs. Working

Group I is probably least involved in the direct support of the CINCs. This is not surprising in light of the comments of the Scientific Advisor to SACEUR cited in the Introduction [1]. However, some of the group's applied efforts are responding to pressing CINC needs. For example, efforts to develop sensor fusion algorithms have found application in many CINC programs (e.g., LOCE). In addition, the group has performed original generic work on measures of effectiveness for command and control systems [11]. This work can be adapted to satisfy the needs of each CINC to evaluate his own command and control system.

Working Group II has been quite active with the component commands and, to a lesser degree, with CINC communities. As an example, DARPA has sponsored a number of research programs to explore distributed C³ concepts. In concert with the Army, they embarked upon the Army-DARPA Distributed Communications and Processing Experiment (ADDCOMPE) to develop and apply packet switching techniques and distributed databases to the needs of the tactical community. Although the effort was focused on the XVIII Airborne Corps, initiatives are underway to transfer the resulting technology to USCENT-COM. DARPA is currently in the initial stages of an analogous program in the strategic arena, the Survivable Adaptive Planning Experiment (SAPE). This program will assist CINCSAC and the Joint Strategic Targeting and Planning Staff by exploring more flexible, adaptable communications and selected decision aids. DCA is also supporting the CINC community through its recently expanded Command Center Improvement Program (CCIP). The CCIP subsumes both fixed and mobile command centers, associated information systems, the standardization and institutionalization of the modular building concept, and a command center laboratory for rapid prototyping and evaluation. The CINCs have benefited substantially from this program, most notably through the MAC model command center and the Proof-of-Concept/Experimental Testbed (POC/ET) for SAC, a derivative of the modular building concept.

Working Group III has had mixed success in supporting the CINCs. In the area of decision support systems, there has been a great deal of activity, frequently in concert with the C^2 systems concepts and components group. However, much of this work is focused on needs of the component commands. For example, one of the major successes in ADD-COMPE was the Automated Loading Planning System (ALPS), which enabled planners to configure complex loadings on transport aircraft more rapidly and with reduced manpower. Also, DARPA is working with the Army to develop prototype corps and division-level planning aids through the Airland Battle Management (ALBM) program. The subset of this group involved with organizational research appears to be less effectively coupled with the CINC community. There are, however, several notable, promising initiatives in this area: the National Defense University is supporting USSOUTHCOM, by exploring alternative organizational concepts for the command and Net Assessment. OSD is employing Carnegie-Mellon University to apply organizational theory to assess elements of USEU-COM's command structure.

Working Group IV has had selected successes in supporting the CINCs. The CINCs are responsible for generating C^2 System Master Plans and contributing to the Global C^3 Assessment. DCA has developed a mission-oriented approach to C^3 planning [12] that is being used by the CINCs and is being adapted to meet the planning needs of Major NATO Commanders [13, 14].

In the area of exercise evaluation, two major successes have emerged under the leadership of DCA. In the strategic arena, they manage the POLO HAT exercises under the sponsorship of JCS [15]. The purpose of these exercises is to measure the effectiveness of C^3 systems that provide two-way connectivity between the National Military Command System and the strategic nuclear forces. Over the years, these efforts have significantly enhanced our understanding of the strategic problem and have stimulated major improvements to the system. In the theater/tactical arena, significant progress is being made by DCA in its development and application of the Headquarters Effectiveness Assessment Tool (HEAT) [16]. It has been employed in several exercises to quantify the performance of the headquarters staffs.

Finally, significant resources have been allo-cated for developing test beds and war games to assist the CINCs in training their staffs and evaluating C^3 performance. Three of the more notable initiatives have been the Warrior Preparation Center, the Joint Warfare Center, and the Enhanced Naval War Gaming System. research is underway to establish the technology to link selected test beds to permit evaluation of distributed C^3 concepts [17].

Challenges for the C^2 Research Community

The current level of support that the C^2 research community provides to the CINCs is quite impressive, even if it is not widely recognized. However, in view of the greatly expanded responsibilities of the CINCs, there are many areas where we can, and should, focus our efforts.

The sub-community involved in the theory of C^3 should strive to develop a foundation upon which the CINCs could base the generation of doctrine, tactics, and plans. Our current state of knowledge in this area can be illustrated by relating it to the field of bridge construction. For thousands of years, mankind built bridges without theoretical underpinings to guide design and implementation. However, once that theoretical foundation was developed, it enabled us to fabricate consistently more cost-effective and reliable structures. We can anticipate far more useful doctrine, tactics, and plans once the theoretical underpinings of C^3 have been developed and applied.

The sub-community involved in new C^2 systems and components should pursue a set of initiatives to address three CINC-unique and common needs. First, every CINC is confronted with the problem of threat perception. This requires the development of a set of tools to assist in the collection process (e.g., tasking available sensors) and the assessment of the collected information. An ancillary requirement is for a means to effectively manage the electromagnetic spectrum. Second, there is a pervasive need for a CINC C^3 system that can survive at appropriate levels of conflict. There are many techniques available to achieve those levels of survivability—including hardening, mobility, and distributed communications—and databases, with suitable adaptive management systems. It is necessary to develop and draw upon these techniques to create a customized, survivable capability that is well matched to the unique needs of each CINC. Finally, there is a heightened awareness of the need to develop trusted computer systems for the CINCs. Although this is a formidable undertaking in its own right, it is compounded by the unique environments. For example, in Korea, the systems must not only deal with security issues but must cope with both English and Hangul. Conversely, in Europe we face other theater-unique problems. Currently, if the FRG provides inputs to selected U.S. systems, the information receives a U.S. classification that may preclude its retransmission back to the FRG. The JCS are sponsoring a multi-level security initiative. As a foundation for this initiative, Tiger Teams are being dispatched to study theater-unique situations. A significant infusion of research activity will be required on security issues if the individual CINC needs are to be understood and addressed effectively.

The sub-community involved in decision support systems and behavioral aspects must ensure that the role of the human is not overlooked. As noted above, the focus of many of the decision support system applications has been at the theater component level. To make this work useful to the CINCs, this sub-community must seek to better understand the political/military/economic environments that are unique to each theater. These differences must be accomodated if decision support systems are to win acceptance.

Many CINCs could profit from an extension of the support that the NDU is providing to USSOUTHCOM. A more sound organizational theory would enhance understanding of preferred arrangements for using U.S. forces in joint and combined operations.

The sub-community involved in testing and evaluating C^3 has several overt challenges. The Scientific Advisor to SACEUR requested that the DSB initiate a task force "...to recommend an overall research and test program to develop the analysis, simulations, experiments, and exercises to allow us to evaluate the impact of C^3 on combat (military missions)." In addition, SACEUR has recently asked DARPA to develop a Distributed Wargaming System (DWS) to support realistic staff training.

This sub-community also has a major challenge in assisting the CINC staffs in developing long-range C^3 plans and architectures. Although the mission-oriented approach provides a useful framework, there is still a significant void in user-friendly implementation tools. There is a particular need for tools that: facilitate the derivation of system deficiencies; assist in formulating cost-constrained C^3 system packages; and support the assessment of the impact of C^3 on mission effectiveness. The CINC staffs also require tools to derive lessons learned from exercises and crises.

Caveats

There are several obstacles to overcome if the C^2 research community is to provide more effective support to the CINCs. These can be cast in the form of actions that the CINCs must take and potential courses of action that the C^2 research community must avoid.

The CINCs must enter into the dialogue by formulating forward-thinking doctrine. The C^2 research community deals with the world of the future and must have an understanding of future concepts of operation and doctrine.

The C^2 research community should be sensitive to three points. First, it must resist the temptation to "push" studies and technologies not well matched to the CINCs' needs and capabilities, or risk user dissatisfaction. Second, it must not forget that U.S. forces are "general purpose" and that some level of standardization is important. It must resist the temptation to develop unique products. Finally, the community must not "end-run" the military services charged with developing and fielding C^3 products. Without the participation of these commands, the penalty can be unsupportable, non-interoperable products.

This last issue poses a major dilemma: How can the support of the military services be elicited without incurring severe time delays and bureaucratic burdens? A mechanism to resolve that problem has recently been proposed by a DSB Task Force on Technology Base Management [18] through the concept of Advanced Technology Transition Demonstrations (ATTDs). The ATTD would be a "proof of principle" demonstration in an operational environment. They projected major roles for the user (operator) as program sponsor and the developer (Systems Command) as program manager. Programmatically, they envisioned a typical program of three years at a total cost of $1O-1OO million in 6.3A funds. They recommended that half of all 6.3A programs (i.e., approximately $1 billion) employ this concept. The C^2 research community should be in the vanguard as this concept is implemented.

Summary

The role of the CINCs has expanded significantly in the areas of the PPBS, the requirements generation process, C^3 development and acquisition, and joint training. Most CINCs lack adequate resources to fully discharge the new responsibilities. Although the C^2 research community currently supports several CINCs, the support has been uneven. It has been successful most often when support is well matched to CINC needs and when performed within institutional channels; that is, performed in cooperation with the supporting military service commands.

Several facets of the support have been disappointing. First, it is perceived that the community has not effectively transferred technology to the CINCs. There is a strong and often accurate perception that sophisticated commercial off-the-shelf products are readily available to the civilian sector, while the CINCs are left with technologically obsolete systems. Second, there is a perception that many of the most exciting fruits of research are overly focused on the needs of the component commands. Finally, there is a sense that the C^2 research community is not adequately attuned to the unique operational context in which the individual CINCs must function. In order to serve that client base effectively, it requires an in-depth understanding of the diverse command issues, force structures, threat environments, and interoperability problems that confront the individual CINCs.

If the research community will recognize and react to these factors, it has an extraordinary opportunity. The most pressing areas involve development of quantitative tools to evaluate the impact of C^3 on mission effectiveness and the development of test beds to support the training of CINC staffs and to facilitate the dialogue between the operational and technical communities.

It is recommended that the JDL consider the utility of C^2 research activities to the CINCs when it is formulating and prioritizing the C^2 research program. It would enhance the likelihood that the products are transferred and used effectively. It would encourage the CINC support and advocacy that is so critical to the survival of the C^2 research program.

References

1. Correspondence from P.J. Berenson, Scientific Advisor to SACEUR, to R. Hillyer, Technical Director, NOSC, June 16,1988.
2. "Policy and Procedures for Management of Command, Control, and Communications Systems," Joint Chiefs of Staff, SM-684-88, August 23, 1988.
3. "Global C^3 Assessment," Joint Chiefs of Staff, June 1986.
4. Deputy Secretary of Defense Memorandum on Revisions to POM Review, Fall 1985.
5. Deputy Secretary of Defense Memorandum on FY88-92 Program Review. Memo DRB 86-4, April 24, 1986.
6. Goldwater-Nichols DOD Reorganization Act of 1986, HR3622, September 30, 1986.
7. "A Quest for Excellence," Blue Ribbon Commission on Defense Management, June 1986.
8. Gen. B. Rogers, SACEUR, Testimony Before the Senate Armed Services Committee, March 12, 1986.
9. Correspondence from P. J. Berenson, Scientific Advisor to SACEUR, to R. R. Everett, Chairman, DSB, February 4, 1988.
10. Report of the DSB C^2 Systems Management Task Force, July 1987.
11. R. Sweet, et al., "Command and Control Evaluation Workshop," Military Operations Research Society, Naval Postgraduate School, January 1985 (Revised June 1986).
12. D. T. Signori and S. H. Starr, "The Mission Oriented Approach to NATO C^2 Planning," *SIGNAL*, September 1987, pp. 119-128.
13. M. Sovereign and G. Hingorani, "Long Term NATO C^3 Planning," *Proceedings of the 1988 JDL C^2 Research Symposium,* June 7-9, 1988.
14. D. Griswold, "Mission-Oriented C^2 Planning in Allied Command Atlantic/Allied Command Channel," *Proceedings of the 1988 JDL C^2 Research Symposium,* June 7-9,1988, pp. 595-600.
15. W. Gorham, "Strategic Nuclear C^3 Test and Evaluation." *Proceedings of the 1988 JDL C^2 Researc h Symposium,* June 7-9 1988.
16. P. Feld, "Supporting the Evolutionary Acquisition of C^2 Systems," *SIGNAL,* June 1988, pp. 211-217.
17. V. Monteleon, "Distributed C^3 Testbed and User Interactions," *Command, Control and Communications Technology Assessment: Conference Proceedings,* MTP-87WOOOOl, July 1987, pp. II-16 - II-22 (not available in public domain).
18. Report of the DSB 1987 Summer Study on Technology Base Management, OUSD(A), December 1987.

4

C³ Systems Research: A Decade of Progress

Harry L. Van Trees

Introduction

I appreciate the opportunity to speak on a topic of vital importance to national security. The title was motivated by the fact that many of the activities at this conference resulted from the 1978 report of the Defense Science Board Task Force on C³ [1] and the actions that DOD took in response to the report. My objective is to look at the progress that has been made in the broad area of C³ systems research since that 1978 report. I accepted the invitation because I knew it would force me to thoroughly review the effort of the research community over the last ten years. It has been a challenging effort.

We begin with the report of the Defense Science Board Task Force on C³ and one of its recommendations: "The DOD should develop a *coordinated program* of *research and testing* on command and control *concepts, design,* and system *performance* to provide the *intellectual base* to guide the evolution of improved command and control systems." [emphasis added]

The guidance was broad and the report did not provide many specific suggestions as to what the committee had in mind. Partially in response to this recommendation, Andy Marshall, Director of Net Assessment in OSD and I (then, Principal Deputy Assistant Secretary C³I, sponsored a conference on "Quantitative Assessment of the Utility of Command and Control Systems" at the National Defense University in January, 1980 [2,3].

The conference was valuable in bringing together diverse interest groups to assess the current state-of-the-art in C² utility (or effectiveness) theory. In 1982, Bob Hermann wrote a report reemphasizing the importance of the problem. Two results that could be partially attributed to the Hermann report were the creation of the Joint Director of Laboratories C³ Research Program and the addition of about two million dollars of funding for C³ research. It should be noted that, starting in 1978, ONR had sponsored a significant research program at Massachusetts Institute of Technology and held an annual Command and Control Workshop [4-12]. In 1987, the Basic Research Group of the Joint Directors of Laboratories held the first C² Research Symposium at the National Defense University [13,14]. The Symposium superseded the annual MIT/ONR Conference and expanded the participation.

In 1987, the Defense Science Board sponsored another study on C³ [16]: again Dr. Buschbaum was the chairman. I want to focus on this recommendation from that report:

"To strengthen the intellectual base for command and control we recommend that a comprehensive program devoted to research on command and control be defined and implemented. The research program should delve into all aspects of command and control, not just the technological aspects. It should form close linkages to the several research and graduate education programs in command and control in service and defense educational institutions and should exploit and foster related research programs in our universities."

Once again, there is an emphasis on developing an intellectual base. A new element is "close linkages to the several research and graduate education programs in command and control in service and defense educational institutions and should exploit and foster related research programs in our [civilian] universities." I strongly believe that this linkage is key to a successful research program.

Clearly, the command and control problem is

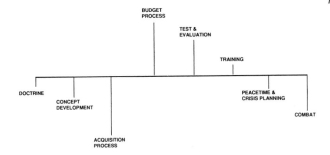

Figure 1. Potential relevance of C³ theory.

central to our national security. Over the years, the demands made on command and control systems have grown exponentially. The increased range, speed, and accuracy of weapons systems have significantly increased the commander's volume of interest and, at the same time, decreased the reaction time. Concurrently, technological developments have provided commanders and their staffs with more capabilities to cope with the C^2 problem. It is well-accepted by both the U.S. and its potential adversaries that C^3 systems are crucial to force effectiveness and there have been numerous examples of problems: for example, the Pueblo, the Liberty, and NORAD computer failures. Finally, C^3 systems are expensive. For example, the cost of each MILSTAR satellite on orbit will be one billion dollars.

While everyone believes that C^3 is important, it is less widely accepted that a theory of C^3 is important. In fact, C^3 theory can have a widespread impact. The range of potential relevance of C^3 theory is shown in Figure 1.

The process of developing C^3 systems begins with concepts and doctrine. More realistic results will be obtained if the C^2 implications of a proposed doctrine, war fighting concept, or weapons systems are considered from the beginning. A good example is the Strategic Defense Initiative. After the initial announcement of the concept, it became clear that battle management and C^3 would play a crucial role. Perhaps if the C^2 implications had been considered earlier, the overall concept would have been different. A major contributor to the acquisition process is an ability to trade-off C^2 system requirements against force effectiveness.

A central feature of the DOD budget process is deciding how to allocate resources among competing demands. For example, should one

spend more money on the Joint Tactical Information Distribution System (JTIDS) to make F-15s more effective in accomplishing their mission, or should one buy more F-15s? Quantitative trade-offs on contributions to force effectiveness could make an important contribution. In the test and evaluation phase, C^2 theory could help define appropriate tests and measures. There is a potential for significant savings in the training area if we could develop realistic C^3 simulations. These simulations could augment field and command post exercises and make them more effective. The potential saving in training costs easily exceeds the entire C^3 systems research budget.

A better understanding of C^2 can aid the process in the peacetime and crisis planning area. Finally, in combat the theoretical results must be applicable in real time. If a theory could generate useful fundamental principles or guidelines, it could augment the commander's experience in making decisions. Decision support systems can also play a useful role. This spectrum provides a rich menu to which the research community can contribute.

Figure 1 portrays the relevance of C^3 theory to a wide range of problems. From the standpoint of a researcher, it is also important to note that C^3 theory is an intellectually exciting area in its infancy. This is in contrast to both communications and control theory which are relatively mature fields (e.g., [17]).

With this as background, I set out to review the progress in the last decade. One of my colleagues remarked that I was spending an inordinate amount of time in preparation: since I had been responsible for C^3I in OSD, I should be familiar with the C^3 research area. The problem is illustrated in Figure 2. The sizes of the areas could represent either money spent or the number of people involved. Most decisions

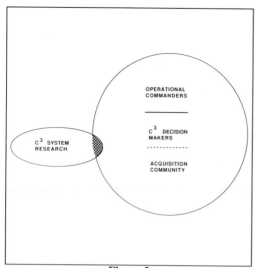

Figure 2.

made by the acquisition community, operational commanders, or C^3 decision makers are made independently of the C^3 systems research community. The basic problem is that the intersection of the sets is almost empty. Very few people are involved in both worlds. This is a fundamental problem. If we are going to be successful, we must get more of the larger community involved in our research.

You cannot study C^2 theory in a vacuum. It is possible to study communications theory in an abstract manner and obtain useful results. You can do research on coding theory or modulation theory without knowing how a modem or a decoder works. I do not think that is true in the C^2 area. I maintain that everyone involved in C^2 research should start by learning some-

thing about warfare. There is not an adequate theory of warfare, but there is good research and excellent discussions. Sun Tsu is an early and worthwhile reference [13,14]. Writing in the 5th Century B.C., he is often referred to as "the father of the theory of strategy." If you adapt his language a little from translated Chinese, you find that many of his ideas are completely relevant to current strategies.

Strategy as we currently know it came into being around 1800 with Napoleon's campaign. Karl von Clausewitz's major work, *On War* [21,22] has had a major impact in many nations. Many of his observations are valid today:

•"Many intelligence reports in war are contradictory; even more are false, and most are uncertain, "

•"Everything in war is very simple, but the simplest thing is difficult. The difficulties accumulate and end by producing a kind of friction that is inconceivable unless one has experienced war. "

•"A critic should never use the results of theory as laws and standards, but only as the soldier does—as aids to judgement."

Earlier, in 1714, Chevalier Folard had introduced the idea of "the fog of war" [23,24] to denote the uncertainty inherent in warfare. Fog and friction still remain major challenges to our C^3 systems. An excellent contemporary reference is Van Crevald's *Command in War* [25]. Other contemporary documents dealing with doctrine and warfighting are FM 100-5, AFM 1-1, and TACM 2-1 [26-29].

OBJECTIVE	Direct every military operation toward a clearly defined, decisive, and attainable objective
OFFENSIVE	Seize, retain, and exploit the initiative
MASS	Concentrate combat power at the decisive place and time
ECONOMY OF FORCE	Allocate minimum essential combat power to secondary efforts
MANEUVER	Place the enemy in a position of disadvantage through the flexible application of combat power
UNITY OF COMMAND	For every objective, ensure unity of effort under one responsible commander
SECURITY	Never permit the enemy to acquire an unexpected advantage
SURPRISE	Strike the enemy at a time or place, or in a manner, for which he is unprepared
SIMPLICITY	Prepare clear, uncomplicated plans and clear, concise orders to ensure thorough understanding

Table 1. Army Principles of War.

THE TWO ASPECTS OF C² SYSTEMS

• **Integrated Systems**
>   doctrine
>   procedures
>   organizational structure
>   personnel
>   equipment
>   facilities
>   communications

• **C² Process**
>   **What does the system do?**

Table 2.

For the Army case, there is a reasonably logical path leading from Clausewitz to FM 100-5. Table 1 is an extract from FM 100-5. Each of these nine principles of war has some C² implication. In particular, the C³ system will play an essential role in the unity of command. The Army uses these principles as a basis for its doctrine, strategy, and tactics, which, in turn, form the basis of requirements for C³ systems.

There are analogous historical bases for the current principles of Air Force and Navy operations. In the Air Force, the heritage leading to the current doctrine comes from Douhet [30], Mitchell [31], and Seversky [32]. (Although some of their theories were not confirmed in World War II). In the Navy, it comes from Mahan [33-35] (Cf.[36] for a summary). The point is that a C² researcher must have a reasonable appreciation for the history and principles of war fighting to set the research context.

With this brief discussion of war fighting we can now continue with the command and control problem.

We begin with the JCS definition of command and control:

"The exercise of authority and direction by a properly designated commander over assigned forces in the accomplishment of his mission. Command and control functions are performed through an arrangement of personnel, equipment, communications, facilities, and procedures which are employed by a commander in planning, directing, coordinating and controlling forces and operations in the accomplishment of his mission." [37]

This definition leads us to the two aspects shown in Table 2. The first is that of an integrated system and represents a physical view. The second aspect focuses on the functional view. What does the system do and how does the process work? It is important that we study the systems from both aspects.

One of the essential elements in understanding C² systems is the evaluation of their capability. Figure 3 shows the spectrum of available evaluation techniques.

The techniques in order of increasing complexity are:
• expert judgement
• analysis
• computer simulation

EVALUATION: SPECTRUM OF TECHNIQUES

TECHNIQUE	COST	LEAD TIME	REPLICABILITY	BREADTH OF APPLICATION	REALISM
COMBAT/CRISIS	N/A	N/A	NONE	LIMITED BREADTH	EXCELLENT
EXERCISES:					
- FIELD - CMD POST	VERY HIGH	2-3 YEARS	LITTLE		GOOD
WARGAMES AND BATTLE SIMULATIONS	HIGH	1-2 YEARS	SOME, WITH PLANNING		GOOD
TEST BEDS	MODERATE		YES	MORE	FAIR
LABORATORY (MAN-IN-LOOP) SIMULATIONS		6 MONTHS TO 1 YEAR		MUCH BROADER	FAIR FOR SELECTED VARIABLES
COMPUTER SIMULATION	LOW		FULLY	VERY BROAD	FAIR
ANALYSIS	LOWEST	WEEKS			

THREAD OF CONSISTENCY

COMMON MEASURES

Figure 3.

- man-in-the-loop simulations
- test beds
- war games/battle simulations
- exercises
- combat

The figure gives a qualitative assessment of how the various techniques compare in cost, lead time, replicability, breadth of application, and realism. The arrow on the left conveys two important points. First, it is essential to have a common set of measures that are applicable and measurable across the spectrum of techniques. Second, these measures must give consistent results so there is a thread of consistency across the techniques. A simple example of this idea is the performance of a digital communication system. Analysis will derive the probability of error as a function of the transmission rate and signal-to-noise ratio. Probability of error is measurable in field tests and combat, and results will be consistent if the analysis was done correctly and the equipment meets specification. In more complex systems, particularly those with humans in the loops, it is generally harder to find consistent measures. I will return to this point repeatedly as we discuss progress in modeling. It is the responsibility of the research community to demonstrate that models are consistent with reality. With these observations about C³ systems as background, we can begin our discussion of progress in modeling.

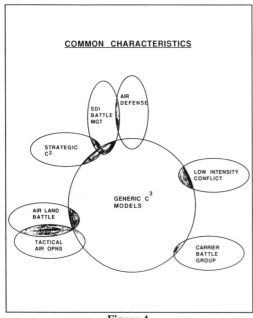

COMMON CHARACTERISTICS

SDI BATTLE MGT

AIR DEFENSE

STRATEGIC C²

LOW INTENSITY CONFLICT

GENERIC C³ MODELS

AIR LAND BATTLE

TACTICAL AIR OPNS

CARRIER BATTLE GROUP

Figure 4.

Progress in Modeling

No single model is going to capture all of the issues: a collection must be employed. The appropriate model will depend on the mission area of interest, as well as other factors. Figure 4 shows that while there is a role for generic C³ models, many of the important features will be mission-specific. The appropriate model also depends on the geographic scope of the model and the level of force aggregation. The choice of a microscopic or macroscopic model will depend on the level of interest. As in any modeling problem, one must match the data available to the degrees of freedom in the model. Once again, it is important to ensure a thread of consistency through the models dealing with a given mission.

Table 3 is a list of useful functional models. Process models focus on the dynamics of the overall system. Often a process model will be embedded with one of the other functional models. The functional models correspond to various tasks carried out in the system. The taxonomy in Table 3 is not unique, but appears to be a useful approach. Some of the functional models will be described subsequently.

To understand a C³ system we must be able to measure its capability. Three types of measures are employed.

- Measures of Performance (MOPs) assess how well a system or subsystem can perform its own internal functions. For example, in the communication system, suitable MOPs include bit error rate, message error rate, end-to-end message delay, antijam margin, and network survivability under various attack scenarios. In a radar system, MOPs include the probability of detection as a function of range, target size, and required probability of false alarm, target resolution capability, and coverage volume.

- Measures of Effectiveness (MOE) assess the capability of the system to enhance the combat mission. One example would be the ability of a JTIDS communications system to improve the kill probability of the F-15 in its air-air mission.

- Measures of Force Effectiveness (MOFE) measure how well the force and the C³ system perform their mission. A typical example would be the damage inflicted on the target set by the ICBM force. The C³ system would influence the result through the sensors, data

C³ FUNCTIONAL MODELS

FUNCTIONAL MODELS	TYPICAL ELEMENTS
• Process Models	System dynamics
• Command and Headquarters Model	Decision-making model, staff elements, data fusion
• Combat and Conflict Model	Physical red-blue engagement model
• Surveillance and Fusion Model	Classical radar, IR, sonar, photo (both fixed, moving), internetting, preliminary fusion. (Red, blue, environment) status reporting, intelligence, I&W, weather forecasting
• Communications Model	Capacity, connectivity, S/N, error rate, security, AJ capability
• EW and Counter C³I Model	Deception, destruction, jamming exploitation
• Information and Control Model	Topology, processing at various nodes, database distribution and management, construction and dissemination of tasking orders, EAMs

Table 3.

processing and display, communication for conferencing, support to decision making, and distribution of the emergency action messages.

MOPs are the easiest to define and calculate and can be obtained in almost all C³ systems and subsystems. MOEs are somewhat more difficult. Sometimes there is ambiguity about whether a particular measure is an MOP or an MOE. The resolution is not vital as long as we understand what is being measured. One of the elements of a comprehensive theory will be an effectiveness theory. MOFEs have been a topic of research for many years (e.g.,[38-40]). The impact of the C³ system adds another degree of complexity to the problem. With these issues and definitions as background, several frameworks for analysis can now be defined. The first is the Mission-oriented Analysis (MOA) approach shown in Figure 5.

Several papers at this conference cover various aspects of MOA (e.g.,[41-44]) so this discussion will be brief. The analysis begins by defining the desired capability objectives for each level of conflict. The levels are normally defined as peacetime, crisis, conventional war, theater nuclear war, and strategic nuclear war. At each level, various degrees of strategic capability are defined. Next we define the missions and the corresponding mission capability levels necessary to accomplish a particular strategic capability. Finally, the components and the corresponding system capability objectives are defined. The C² system (either existing, programmed, proposed, or hypothesized) is then evaluated and the results are passed up to the original level.

Two points concerning MOA are useful:

• The quantitative evaluation at the subsystem level is classical analysis of MOP-type measures. The difficulty arises as one aggregates results to move up the chain.

• The dimensionality increases rapidly and

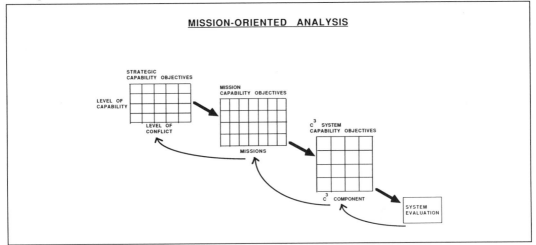

Figure 5.

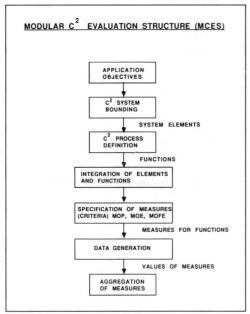

Figure 6.

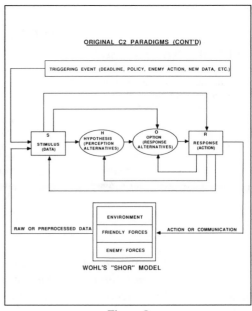

Figure 8.

the analyst must be skilled in recognizing the important cases.

The Defense Communications Agency and the JCS have achieved some useful results from this technique.

The Modular C2 Evaluation Structure (MCES) shown in Figure 6 is a second framework. MCES is described in [45] and is discussed and applied in [46-49]. MCES can be considered a check list to guide the analyst in defining and solving the problem. As with the MOA approach, the key to success is in the skill of the analyst and the relation of the

detailed model to the actual system.

The final technique is the System Effectiveness Analysis (SEA) technique developed by Bouthonnier and Levis [50-51]. They construct a multidimensional space whose axes correspond to various effectiveness measures. MOPs of the systems are then mapped into loci in the effectiveness space. This procedure is more of an analysis technique than it is a framework and could be used with either of the previous frameworks.

The first class of models are process models. I will briefly review some of the classical

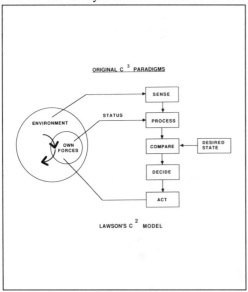

Figure 7.

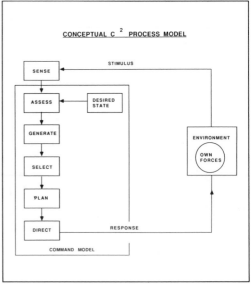

Figure 9.

paradigms and then look at some representative process models. The first model of interest is Lawson's classical model in Figure 7, first published in 1978 [52] and discussed in papers [53,54].

The classic Wohl's SHOR paradigm on Air Force tactical air operations is shown in Figure 8 [55,56]. Both of these models can be mapped into the model shown in Figure 9 [57].

The first box is the sense function where data are collected to describe the current and projected situation. We are specifically interested in the status of enemy forces and their possible actions; status of friendly forces; and the status of environment (for instance, weather, geographical considerations, possible nuclear effects, political constraints, and neutrals). In carrying out this function, we employ a myriad of different sensors, communication, and data fusion systems. In the assess function, the data from the sense function are used to hypothesize the capabilities and possible intention of the enemy; to review the status and capabilities of friendly forces; and to compare the situation to the "desired state." The term "desired state" comes from control theory rather than the usual military lexicon. It can range from a tactical objective such as taking a hill or bridge to a major strategic state. The

next step is to generate options or alternative courses of action to get the desired state. The next block includes evaluation of the options and choosing a course of action. In the planning step, the necessary implementation details are developed to carry out the selected courses of action. Finally, the direction block indicates the dissemination of orders to the appropriate forces. This command model interacts with the environment and the friendly forces in the manner of a closed loop. Over the years a number of alternative paradigms have been developed (for example, Mayk and Rubin [58,59] or Nichols [60]). In general, they include the same functions relabeled or repackaged .

One of the deficiencies in these paradigms is that they do not give adequate emphasis to the enemy. The enemy is not just an element in the environment, but an adversary who is deliberately trying to achieve his own objectives and deny us ours. To emphasize the two-sided nature of the problem, I prefer the model shown in Figure 10 which I originally introduced at the Quantitative Evaluation Conference in 1980[2].

The command section of the previous paradigms are embedded in the boxes, Blue Command and Red Command. The other ele-

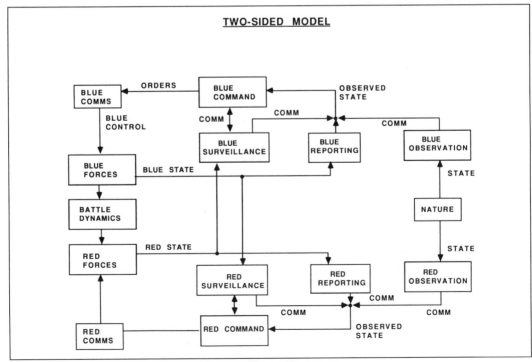

Figure 10.

CONCEPTUAL C^2 TIMELINE

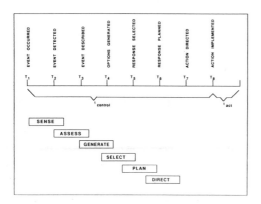

Figure 11.

ments of the C^2 process are specifically identified and can be associated with the model taxonomy in Table 3. Most importantly, the fundamental two-sided nature of combat is made explicit. In all cases, the models must be replicated (or nested) vertically to show the command hierarchy and horizontally to show adjacent units.

The significance of the results in the analysis will depend on how well the analyst treats the various elements of the boxes. In a given application, one should choose the model that best illuminates the important features of the situation at hand. At the current stage of C^2 theory development a standard or generic model is not needed.

The other way that one should view the C^2

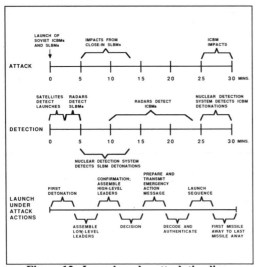

Figure 12. Launch under attack timeline.

process is the classical time line approach shown in Figure 11 (from [45]). It displays the functions from the model in Figure 9 in a time line sequence. An interesting application of this approach is shown in Figure 12 (from [61]). The figure illustrates a possible time line for a Soviet missile attack against the United States and the actions required to respond with a Launch Under Attack option. There is a very small window of opportunity in which the National Command Authorities must make a decision if this option is to be successful. Time lines are an important element in most C^2 analyses. There are useful discussions of timeliness in [62] and [63].

These are some representative dynamic models:

• Classical state variable
  • Thermodynamic (Lawson)
  • Markov (Rubin, Mayk)
• Statistical Mechanics (Ingber)
• Possibilistic (Zadeh, Goodman)
• Catastrophe and Chaos (Dockery, Woodcock)
• Adaptive control (Strack)

Most of them have been discussed at the MIT/ONR or JDL conferences so I will comment only briefly. Classical state variable models have a natural appeal because of their success in treating analogous problems in control theory. Lawson [53] used a thermodynamic analogy to develop a simplified model. Rubin and Mayk [64-66] have treated several interesting cases using a stochastic model based on a discrete time, discrete state Markov process model. They divide the battle into a sequence of stages which are then subdivided into phases. By numerically solving recurrence equations, they obtain probability distributions for various force effectiveness measures.

Ingber [67-69] has used a statistical mechanical approach to derive dynamic models. The central feature of his stochastic model is the path-integral Lagrangian. In his paper at this conference [70], he discusses an approach which is consistent with the theme on models that I will emphasize later. He is running data from the National Training Center and comparing the results with outcomes at NTC.

Goodman [71-73] has developed a general model for C^2 processes by viewing them as interacting networks of node complexes of decision makers. The nodes represent a wide variety of entities, that is, troops, tanks, ships, fusion centers, and command centers). The node state and structure, its input signal variables, and its output signal variables (for all the nodes in both the friendly and enemy system) are used to characterize the processes. In order to characterize the evolution of the state of each of the nodes, he introduces various logical relationships (for example, classical Boolean logic, probability logic, Zadeh's fuzzy logic, or Dempster-Shaefer belief logic). With these ideas as a basis, he formulates a formal procedure for evaluating the evolution of the system.

Dockery and Woodcock have developed models using catastrophe and chaos theory [74-77]. Chaos theory [78] deals with physical phenomena in which nonlinear interactions cause small changes in initial conditions to lead to drastically different system behavior.

Strack [79] models the system as an adaptive control system and uses regression analysis to develop effectiveness.

At their present stage of development it is premature to attempt a comparative analysis of the value of these various models. To make these or other dynamic models useful, three steps are important:

• Model a reasonably realistic scenario,

• Correlate the results with a simulation, test bed, or actual experience and understand the differences between the results.

• Explain the lessons learned and the significance of the conclusions in terminology that is familiar to a commander.

The last step is particularly important if C² research is ever to have an impact on the real world.

Command theory is the first functional area of interest. Some of the representative elements of a command theory are:

• What decisions are going to be made at each level in the hierarchy?

• What constraints are there on the commanders options?

• With whom does the decision maker have to interact ?

• What information is required to make each decision?

• What is the decision time line?

• How timely, accurate, and complete must the input information be?

• How should the information be presented to the commander?

• What decision aids are needed?

• What is human behavior in a stressed environment?

Cushman [80] has identified some of these elements in "A Commander's Catechism." The first element identifies the entire structure of the C² process: Who is authorized to make what decisions? The other elements delineate other issues that must be considered in the command function. The list is representative, not exhaustive.

Representative elements of a headquarters theory are:

• Information flow patterns;

• Intra-nodal processing, communications, and displays;

• Physical topology of headquarters function (distributed, dispersed, centralized);

• Database structure and maintenance;

• Decision aids, the use of artificial intelligence; and

• Survivability of headquarters function

The headquarters functions are those processes needed to support the commander(s). Although both the command and headquarters functions have been carried out for several centuries, often with spectacular success, the development of adequate theories is still an area of active research. In many cases, the successes appear to have been due to applying Clausewitz's dictum: "Always have a genius in charge, first in general and then at decisive points" [21].

Some representative command and headquarters models are:

• Headquarters effectiveness assessment tool (HEAT),

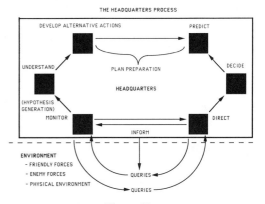

Figure 13.

• Data flow and decision making structures (Petri Nets),
   • Models of decision makers, and
   • Resource allocation.

The Headquarters Effectiveness Assessment Tool (HEAT) was originally developed by Hayes and is being developed further by various researchers (e.g., [81]). The HEAT model is shown in Figure 13. It was used originally for headquarters that were primarily responsible for planning, supporting, and coordinating fighting forces. It treats the C^2 process as an information management system and attempts to measure effectiveness in terms of military mission accomplishment. Research is being carried on with generalizations of HEAT.

Some useful work has been accomplished in modeling data flow and decision making structures using Petri nets. Petri nets, which are widely used in computer science (e.g., [82,83]), appear to have first been applied to C^2 problems by Tabak and Levis [84-85]. These papers and subsequent papers [86-90] have used this technique to develop a structured approach to the design of command and control organizations. They first generate a data flow structure using the Petri net formalism. The various types of activities such as data fusion, situation assessment, and response selection can be represented in this formalism. Various MOPs such as accuracy, timeliness, task processing rate, and workload are evaluated. The data flow structures are then transformed into organizational designs that allocate functions to various decision makers. MOEs are then evaluated for the various organizations.

One of the objectives of the research on command and headquarters theory should be to understand the elements of each. Through understanding, we can improve the perfor-

mance of the commander and his staff. We should note that there is a large body of work on decision support systems that we have not covered.

The next category of models are combat and conflict models:
   • Lanchester-type equations (Deterministic differential equations),
   • Stochastic combat models (Markov processes), and
   • Game theory.

The earliest work in this area was by Lanchester [38]. A simple form of his equations for aimed fire are shown in Figure 14. The variables F and E are the sizes of the friendly and enemy forces, respectively, and C_E and C_F are the kills/sec/unit of force. In these simple equations, the impact of C^2 appears in these coefficients. For example, improved detection and resolution capability and better target location accuracy would raise C_E. Delay in receiving target location information for moving targets would decrease C_E. There have been numerous generalizations of Lanchester's equations. Taylor [95] provides a comprehensive overview (without any C^2 emphasis). Moose and Wozencraft [96] presented a useful generalization at an earlier symposium. (See [97] also). Shubik's *Mathematics of Conflict* [98] contains several relevant papers. The various models attempt to incorporate the effects of resupply, self-attrition, heterogeneous forces, and information uncertainty. In order to incorporate the effects of maneuver, a set of coupled partial differential equations may be written [99]. Unlike Lanchester's original equations which could be solved explicitly, most of the generalizations require a numerical solution.

An obvious generalization is to incorporate the stochastic nature of the combat process to obtain a Markov process model. Suitable references include [100-103]. The behavior of the mean value of the process relates back to a version of generalized Lanchester's equations. Game theory also has been employed to model combat (e.g., [104-105]).

Communications theory is perhaps the most advanced of the various functional areas. Representative elements of communication theory are:
   • User requirements (connectivity, quality, capacity, survivability, environment)
   • Generic properties of media and systems

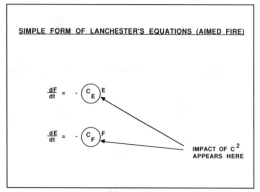

SIMPLE FORM OF LANCHESTER'S EQUATIONS (AIMED FIRE)

$$\frac{dF}{dt} = -\left(C_E\right)E$$

$$\frac{dE}{dt} = -\left(C_F\right)F$$

IMPACT OF C^2 APPEARS HERE

Figure 14.

- Capacity
- Quality
- AJ performance
- LPI performance
- Survivability
- Reliability
- Flexibility
- Connectivity
- Time delays

Starting with the user requirements, the general design procedures are well understood and in most cases the generic measures listed above can be evaluated. There are a large number of classic papers and texts that consider these issues(Cf. [106-109]). Communications is an area in which the research community has had a major impact on procurement of DOD systems. As examples, the ARPAnet significantly influenced current DOD communications networks as Lincoln Laboratories satellite research has influenced the current military satellite architecture.

We should note that all of the above measures are MOPs. An important research area is assessing what impact a given communications MOP has on mission effectiveness. Considerable work is still needed in that area.

Representative elements of a surveillance and fusion theory are:

- Volumes of responsibility and interest
- Generic properties of sensors (e.g., detection, tracking, capacity, processing capabilities)
- Tasking procedures (e.g., responsibilities, timeliness)
- Information flow from non-organic sensors
- Topology of sensor information flow and fusion

A simple definition of the commander's volume of interest is that space around the commander where actions could have a "real-time" or "near-term" impact on mission accomplishment This volume has grown continuously over the years as the range and accuracy of weapons systems has increased. Napoleon's volume of interest was probably a circle with a radius of about 100 miles. The JCS volume of interest consists of the entire globe including the ocean depths and space up to at least the geosynchronous satellite orbit altitude. Sensors have continually tried to keep up with this vol-·ume of interest in order to reduce our uncertainty about the enemy's intentions. New weapons systems such as stealth cruise missiles

cause the race to continue. The volume of responsibility is generally a smaller space and must be clearly defined when the C^2 problem is structured.

The generic properties of sensors and their MOPs are reasonably well understood. As in the communications area, there are a number of classic papers and texts (Cf. [110-112]) that explain how to design radar and sonar systems and measure their performance.   Translating these MOPs to MOEs is still a challenging problem, but a number of useful results have been obtained. Tasking procedures for the sensor systems connect back to the command structure problem.

The state of fusion theory is far less mature. There is a separate panel under the JDL, the Data Fusion Panel, to deal with this area (Cf. [113]). Data fusion remains a major challenge to the C^3 research community.

The two remaining functional models in our taxonomy are the EW and Counter-C^3I Model and the Information Model. In the interest of brevity, we will not discuss them in detail. However, they are both significant areas and should not be neglected in the research program.

The final point to make with respect to the model taxonomy is that there are significant research problems in each of the areas. The C^3 research program should be balanced in its efforts.

Progress in Simulation, Test Beds, and Exercises

The other area of interest is simulations, test beds, and exercises. They deserve equal attention with the modeling area for several reasons. First, as indicated in Figure 3, they are closer to the real world. Secondly, they are much more expensive: DOD probably spends 2-3 orders of magnitude more money in this area than on modeling. Finally, they tend to have more impact on operational commanders and decision makers, so improving their fidelity is important.

However, in the interest of time and space, we will list only some representative facilities for simulation (Table 4), test beds (Table 5), and field exercises (Table 6).

Once again we emphasize the importance of a thread of consistency across the various areas.

REPRESENTATIVE SIMULATION FACILITIES

- NAVY WAR COLLEGE: WARS, NWGS
- TACTICAL AIR WARFARE CENTER: BLUE FLAG
- ARMY WAR COLLEGE: JOINT THEATER LEVEL SIMULATION
- USAFE, RAMSTEIN: WARRIOR PREPARATION CENTER
- IFFN JOINT TEST AND EVALUATION
- NOSC: IBGTT/RESA
- MAXWELL FIELD: COMMAND READINESS EXERCISE SYSTEM
- JCS: JESS
- JDL - SIMNET
- NPG: WARGAME LAB

Table 4.

REPRESENTATIVE TEST BEDS

- **NATIONAL TRAINING CENTER**
- **RED, GREEN FLAG**
- **9TH INFANTRY DIVISION, FT. LEWIS**
- **ADDCOMPE, FT. BRAGG**
- **AEGIS TEST BED, MOORESTOWN, NJ**
- **SDI NATIONAL TEST BED**

Table 5.

REPRESENTATIVE FIELD EXERCISES

- **GALLANT EAGLE**
- **NORTHERN WEDDING**
- **REFORGER**
- **SOLID SHIELD**
- **GLOBAL SHIELD**

Table 6.

Fundamental Principles

After completing this review of progress, a logical question is: Are there any fundamental principles? If one is looking for something as fundamental as Shannon's Channel Capacity Theorem [114] or Heisenburg's Uncertainty Principle, the answer is clearly no. If one is looking for important central results such as optimum Wiener filters, Kalman filters, the fundamental importance of eigenvalues and eigenfunctions, or ambiguity functions, (Cf. [109,110]) the answer is probably still no.

However, four notions seem to frequently arise:

• Uncertainty is fundamental in warfare. As we get more information (versus data) our uncertainty decreases. As time increases the uncertainty will normally approach some minimum value. Unfortunately, the window of opportunity may close before the uncertainty reaches an acceptable level. Thus the commander and his staff must deal with this time—uncertainty tradeoff. It is important for the staff to apprise the commanders of the specific uncertainties that are inherent in the information they are presenting.

• Combat and associated C^2 is inherently a two-sided problem. At the system design level, we must remember that the communications and radar systems must work in a hostile environment created by both the enemy and nature. The ability to function in the presence of jamming, physical attack, EMP, and other threats must be an important design factor. At the decision making level, we must remember that we are dealing with an intelligent adversary and consider his possible reactions to our strategies. We also must remember that we are fighting an information war and that disrupting, confusing, or destroying his C^3 is a key element in an overall strategy.

• We must be able to react faster than the enemy. This principle applies from one-on-one air engagements to entire campaigns. History is replete with examples, from Alexander, Caesar, and Napolean to Patton and MacArthur, where reaction times were important to battlefield success. The Son Tay prison camp raid in 1970 is an example of a decision/implementation cycle that lead to failure. It took about six months from the time the location was first identified until the raid was conducted, only to find the prisoners gone. Every element of the C^2 process must recognize the importance of timely decisions.

• The final principle is that force multipliers can start at zero. We often refer to C^2 as a force multiplier. Basically, if we take as a reference point two opposing forces and strategies, with assumed C³ systems, then there would be a given battle outcome. This outcome (or MOFE) can be measured in different ways (e.g., surviving forces, territory occupied, objectives obtained). By examining the actual C³ system a given MOFE will be changed. Thus,

$$MOFE_N = \alpha MOFE_R$$

where the subscripts N and R denote "new" and "reference," respectively, α is the C² force (effectiveness) multiplier. In many cases, one can show that improved C² can significantly increase force effectiveness. Unfortunately, it also can significantly reduce effectiveness. The capture of the Pueblo is an example where inability to summon timely air cover resulted in the force effectiveness being reduced to zero.

There are probably other equally important notions as these four principles. Since I was unable to develop any really fundamental principles, it will be useful to discuss what I will immodestly refer to as Van Trees' Laws of C³.

Van Trees' Laws of C³

Law 1: Budget
The first law deals with the budget and is motivated by Figure 15. C³ was 4 percent of the total defense budget in 1970. By 1985, it had grown to 7 percent. Using linear extrapolation we obtain the first law:
• "By 2048, the entire defense budget will be allocated to C³."
The first corollary in this law is that:
• The C² force multiplier will have to approach infinity, because there will be no forces left.

It is important to note that C³ budget figures do not include intelligence systems. If intelligence were included, the result would be even more dramatic.

A second corollary follows if one plots the cost of a communications satellite (on-orbit) starting with the DSCS-I (IDCSP) in the 1960's and concluding with the projected MILSTAR satellite in the early 1990's. Then compare those costs with the total C³ budget. Although we do not show the curve, it results in the second corollary:

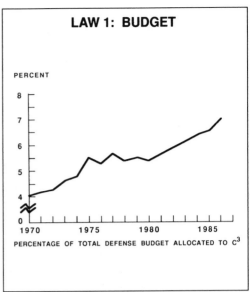

Figure 15.

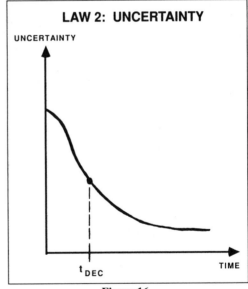

Figure 16.

• By approximately 2032, the entire C³ budget will be devoted to a single military communications satellite.

The positive side of this corollary is that it should reduce interoperability problems and simplify the C³ budgeting process.

While the first law states an improbable conclusion, it should demonstrate the importance of understanding the utility and effectiveness of C³ systems, so we can determine how much is needed.

Law 2: Uncertainty

The second law deals with uncertainty and is illustrated in Figure 16.

The sketch is a plot of uncertainty using an appropriate measure as a function of time. As more information about the situation is assimilated the uncertainty decreases. Unfortunately, the second law says:

"The uncertainty in a C^2 system generally reaches an acceptable level after the required action time is past"

Thus, the commander must work with the uncertain information that is available at the time. The strategic C^2 example in Figure 12 is a dramatic example, because the NCA must make a LUA option decision within about 15 minutes after a Soviet launch. At that time there may still be significant uncertainty about the situation. Examples of C^2 decisions throughout history show similar phenomena. Technology is not going to change this basic relationship. For example, as sensor ranges go up, the volume of interest will increase. The successful commander must cope with these uncertainties (the fog of war) and make appropriate decisions. The role of the C^2 system is to reduce the uncertainties as much as possible and make certain the commander understands them.

Law 3: Communications Capacity

The third law deals with the capacity of the communications systems and is shown in Figure 17.

If we plot the communications capacity in bits/sec as a function of time, it is easy to show exponential growth. A good example is the transatlantic military communications capacity in World War I, World War II, 1965, and now. The exact current capacity may be classified, but by counting satellites and cables, we can see that it must be in the range of 500 mbps to 1 gbps, compared to a miniscule World War I capability. Unfortunately, the user demand has grown even more dramatically, leading to the third law:

"The total communication capacity of C^3 systems has grown as $A_o e^{\alpha t}$. The perceived user requirement has grown as $B_1 e^{\beta t}$, where $\beta > \alpha$."

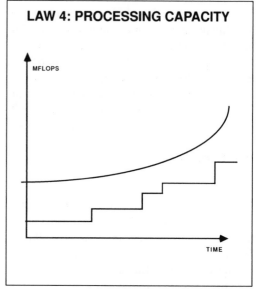

LAW 4: PROCESSING CAPACITY

MFLOPS

TIME

Figure 18.

In all of the architecture studies, one sees an insatiable demand for more capacity. One reason is that, unlike the civilian community, the actual military user or the person determining communications requirements, does pay the telephone bill. Thus, the tradeoff between required capacity and cost is less direct. In spite of this, the community must develop methods of determining how much is enough, in terms of the contribution of communications capacity to force effectiveness.

Law 4: Processing Capacity

A similar phenomenon for information processing capacity is shown in Figure 18. As

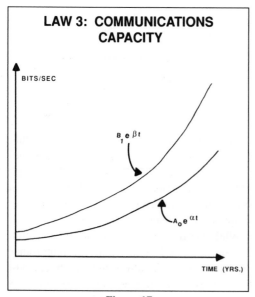

LAW 3: COMMUNICATIONS CAPACITY

BITS/SEC

$B_1 e^{\beta t}$

$A_o e^{\alpha t}$

TIME (YRS.)

Figure 17.

each new generation of computers has been introduced into command centers the information processing capacity has increased dramatically. A curve fitted to the growth would show exponential growth as a function of time. However, the fourth law states:

"The information processing capacity of computers in command centers has grown exponentially. The perceived processing load has grown by a larger exponent."

Examples of this phenomenon abound. For example, in Vietnam during 1966, there were 500,000 messages per month coming out of the two communication centers. This corresponded to an enormous information processing load. When you look at the results of JCS exercises or processing loads during crises, you see exactly the same behavior. A significant challenge to the research and operational community is to match the information processing demands to what the commander needs and his staff can absorb.

Law 5: Duality

The next law deals with what I call "duality." A common approach among mathematicians and researchers is to translate a new problem into one that someone has already solved and then apply the earlier result. One can construct reasonably good analogies between C^3 and living (biological) systems. One can also construct analogies between C^3 systems and the management of large companies. Unfortunately, these analogies lead to law 5:

"The C^2 problem is analogous to the living system problem. By mapping one into the other, we can replace one unsolved problem with another unsolved problem."

The corollary 5A is that you can map the C^2 problem into a business management problem that is equally hard to solve. We would probably hesitate to use U.S. Steel, International Harvester, or Texas Air as models of how to run the military. The point of this law is that analogies are interesting, but duality is only useful if it helps obtain quantitative results or intuitive understanding.

Law 6: Acronyms

The next law deals with acronyms and is related to Augustine's Ninth Law [115] which is:

"Acronyms and abbreviations should be used

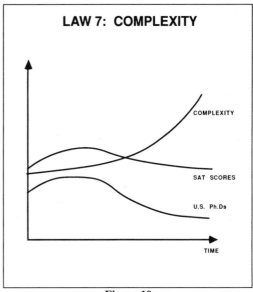

Figure 19.

to the maximum extent possible to make trivial ideas profound."

This general law refers to the entire Defense Department. However, the C^3 community has developed "acronymization" to a fine art. The result is Law 6:

"If one takes the 16 most frequently used letters in the English alphabet and creates an ordered triplet randomly from this set, the probability is 0.246 that the result will be a segment of a C^3 acronym."

For example, if the triplet comes out SBR, one has the space-based radar; TID gives JTIDS; and so forth. Note that by using .246 instead of .25 in the result, we have added credibility to the conclusion. Acronyms are just a visible sign of our difficulty in communicating with the rest of the world. Overcoming this communications problem is an important challenge.

Law 7: Complexity

As everyone realizes, C^3 equipment is becoming more complex with each new generation, as shown in Figure 19.

We also show two metrics relevant to dealing with the population that must develop the basic theory, design the systems, and operate and maintain the equipment. Although the decline in SAT scores appears to have leveled off, the statistics concerning math and science capabilities of high school and college graduates is alarming. It should be noted that the military services have done an excellent job in raising the average education level of their recruits. A related and equally disturbing statistic is the decline in Ph. D.s in math, science, and engineering awarded to U.S. citizens. This trend affects not only our C^3 capability, but also the ability of our nation to compete in the world market. This complexity-ability divergence increases the importance of equipment reliability, built-in test equipment and extensive training under realistic conditions. It also emphasizes the importance of reducing equipment complexity by keeping the specifications reasonable. Often the last 0.5dB of performance increases the complexity (and cost) significantly.

Law 8: Models versus Reality

C^3 researchers are often criticized because their models do not correspond to reality. Actually the problem may rest with operational commanders who can not seem to understand that:

A. To move a battalion 100 miles is a simple stroke with the "mouse"

B. A flanking attack is accomplished by transposing the transition matrix

The research community knows that if you portray the corps sector and the location of its units on a computer screen, then you can move a battalion by simply moving the "mouse. " Similarly, if you represent the motion of a particular unit by an appropriate Markov process, then you can conduct a flanking attack by a matrix transposition. Unfortunately, operational commanders feel that the actual operations are not so easy.

This law relates back to one of the important points in the earlier Progress in Modeling discussions. In order for models to be useful they must correspond to a realistic scenario; correlate with simulations or test beds or actual experience; and be explainable to the operational commander, in his terms. The onus for accomplishing these steps rests on the model maker, not the operational commander.

Law 9: Predicting Future C^3 System Performance

The ninth law deals with predicting the performance of C^3 systems as follows:

"The performance of future C^3 systems generally turns 'green' at T_o, where T_o is greater than:

(1) End of current administration plus 8 years.

(2) Proponent's retirement age plus 3 years."

Representative mission-oriented analyses typically evaluate the current system, the near-term system and a far-term system. The normal effect from using analyses with red, yellow, and green as metrics is that, in the current system, a fair number of missions are red and yellow. From this we create packages or programs to improve the posture. By the near-term you will still have a mixture of red and yellows, but with more greens. There is a reasonably general phenomena that almost all missions turn green in the far-term, particularly if one is willing to supply the funds that the proponent is requesting. It is most interesting to note the point at which everything turns green. What the data show is that T_o is generally one of two times. It can be the end of the current administration plus 8 years. In this option it gives the system time to circulate through two new sets of political appointees. Therefore, if it does not actually turn green, the current person has at least one or two preceding generations to blame. Alternatively, from the standpoint of the proponent: it is generally his retirement age plus about 3 years.

Unfortunately, past experience indicates the actual performance of future C^3 systems is generally not as "green" as projected in their early stages of planning. Among other factors, Laws 3 and 4 contribute to this problem.

Law 10: Interoperability

One version of this law is:

"When Clausewitz talked about 'friction in war,' he was predicting the problems with C^3 in joint operations."

Lack of C^3 system interoperability is one of the key problems. All this law does is point out that the phenomena was predicted a number of years ago. It is still one of the most significant problems that we have in the C^3 world and deserves the attention of the research community.

Although there are a number of other laws dealing with such interesting phenomena as sensor range versus volumes of interest, the ratio of signal personnel to the size of the total force, the effects of compartmentation, the characteristics of software, and the size of staffs, it is appropriate to stop at ten. One should also note that a number of Augustine's Laws [115: Laws 5,12,15,17,18,19,37, and 43] can be applied directly to C^3 systems.

While I presented these laws in a somewhat humorous vein, they more than hint at phenomena that must be recognized in research.

Problems and Shortfalls

We have not done an adequate job of relating our research to actual C^3 operational problems, that is, to the real world. It goes back to the diagram in Figure 2 There are two worlds with very little overlap: commanders and decision makers with real problems and our small research community. The onus is on us to increase that overlap. It is important that we be able to tell the operational and decision making communities what our research is good for: to tell how this research will help them understand and solve C^2 problems.

The DSB Task Force said similar things in 1978 and 1987. The implicit undertone is that not really much had been accomplished. An important reason for this attitude is that we, as a research community, have not done a good job of articulating our accomplishments. If you look at the work that is presented here and the work that has been presented in some of the earlier conferences, we have actually made a fair amount of progress in understanding parts of the problem. While we do have a very long way to go, we have not done a good job of articulating the results and showing how they apply to C^2 problems.

As a discipline, C^2 research is not yet over the academic threshold. Look at many of the people in the universities who are working on C^2 research. They were already established in their own fields, such as control theory, communication theory, or systems theory, before they became involved in C^2 research. It is vital that the C^2 research field have academic respectability if we want to create a body of researchers in various universities. They must have the same stature as the other disciplines. To achieve this credibility, we need journals with real peer reviews. We need some segment of a professional society that is interested in C^2 research as its major thrust. It is very important that we show young professors a career path that can lead to tenure by becoming known as an expert or qualified authority in this field. This is not a uniform situation: there are several universities such as M.I.T., Carnegie Mellon, Connecticut, Naval Postgraduate School, and George Mason where C^2 is recognized as a legitimate and intellectually exciting discipline. This academic involvement is essential to developing a C^2 theory.

We have not done an adequate job of incorporating the human element (the commander, staff, and other involved people) into our analyses. It is vital that we create adequate models or utilize test beds that provide a realistic interface with the people involved, otherwise we are not going to have valid results.

We do not have adequate interaction with the classified community. I would expect that 70 to 80 percent of this audience is cleared for some access to classified information. There is a fair amount of interesting work going on behind that curtain and we have to find a mechanism to participate. One approach is to have a classified session as part of these conferences.

Finally, I think we need to establish broader interaction with the operations research community. The OR community has at least a fifty year history of modeling and analyses in support of general military operations. While they have not emphasized C^2, they have developed a number of interesting techniques, mathematical models, and simulations that may be readily adaptable to our problems

Summary

C^3 systems theory, however it is defined, is still in its early stages. I believe that significant progress has been made since 1978, but, in spite of the progress (which is partially reflected in the references I have included), we must do a better job in communicating with the real world.

Acknowledgments

In the course of research for this paper I asked a number of people for input. The following were kind enough to respond:

S. Andriole, M. Athans, E. Brady, A. Campen, G. Coe, F. Diamond, J. Dockery, I. Goodman, L. Ingber, J. Lawson, D. Leedom,

A. Levis, M. Muderian, E. Rechtin, R. Robinson, R. Sabat, M. Sovereign, C. Smith, R. Sweet, S. Starr, and J. Wohl.

References

[1] Office of the Undersecretary of Defense Research and Engineering,"Report of the Defense Science Board Task Force on Command and Control Systems Management," July 1978.

[2] Van Trees,H.L.,"Keynote Address", Proceedings for Quantitative Assessment of the Utility of Command and Control Systems, National Defense University, Report MTR-80W00025, January 1980.

[3] *Proceedings for Quantitative Assessment of Utility of Command and Control Systems*, National Defense University, Washington,D.C., Report MTR-80W00025, January 1980.

[4] Athans,M.(ed.), Proceedings of the First MIT/ESL-ONR Workshop on Naval C3 Systems, ESL-R-844-846,M.I.T., Cambridge, MA, Sept. 1978.

[5] Athans,M.(ed.) Proceedings of the Second MIT/ONR Workshop on Naval C3 Systems, LIDS-R-966-968, M.I.T., Cambridge, MA., March 1980.

[6] Athans,M., Ducot, E., and Tenney, R, (eds.) Proceedings of the Third MIT/ONR Workshop on C3 Problems, LIDS-R-1020-1023, M.I.T., Cambridge, MA, Sept. 1980.

[7]Athans,M., Ducot,E., and Tenney,R., Proceedings of the Fourth MIT/ONR Workshop on C3 problems, LIDS-R-1156-1159, M.I.T., Cambridge, MA, October 1981.

[8] Athans,M., Ducot,E., Levis,A.,and Tenney,R.(eds.) Proceedings of the Fifth MIT/ONR Workshop on C3 Systems, LIDS-R-1267, M.I.T., Cambridge, MA., December 1982.

[9] Athans M., Ducot,E., Levis, A.,and Tenney R.(eds.) Proceedings of the Sixth MIT/ONR Workshop on C3 Systems, LIDS-R-1354, M.I.T., Cambridge, MA., December 1983.

[10] Athans,M. and Levis,A. (eds.) Proceedings of the Seventh MIT/ONR Workshop on C3 Systems, LIDS-R-1437, M.I.T., Cambridge, MA, December 1984.

[11] Athans,M., and Levis, A. (eds.) Proceedings of the Eighth MIT/ONR Workshop on C3 Systems, LIDS-R-1519, M.I.T., Cambridge, MA, December 1985.

[12] Athans, M. and Levis, A., (eds.) Proceedings of the Ninth MIT/ONR Workshop on C3 Systems, LIDS-R-1624, M.I.T., Cambridge, MA, December 1986.

[13] Proceedings: Command and Control Research Symposium, JDL Basic Research Group, National Defense University, 1987.

[14] Johnson, S. and A. Levis (eds), Science of Command and Control: Coping with Uncertainty, AFCEA International Press, Washington, D.C., 1988.

[15] Proceedings of the JDL BRG C Symposium, Monterey, CA,June,1988.

[16] Office of the Undersecretary of Defense for Acquisition, "Report of the Defense Science Board Task Force on Command and Control Systems Management," July 1987.

[17] Athans, M., "Command and Control (C2) Theory: A Challenge to Control Science ," IEEE Transactions on Automatic Control, Vol. AC-32, No. 4, April 1987.

[18] Sun Tsu, The Art of War, edited by James Clavell, Delacorte Press, New York, NY, 1983.

[19] Sun Tzu, The Art of War, translated by Samuel B. Griffin, Oxford University Press, New York, 1971.

[20] Phillips, T.R. (ed.), Roots of Strategy, Military

Service Publishing Co., Harrisburg, PA, 1940.

[21] Howard, M. and Paret, P. (eds), Karl von Clauswitz, On War, Princeton University Press, Princeton, NJ, 1976.

[22] von Clausewitz, K., On War, translated by O.S.M. Jolles, Infantry Journal Press, Washington, DC, 1950.

[23] Folard, Chevalier, Nouvelles Decouvertes sur la Guerre, 1724, quoted in Heinl, R.D., Jr.'s, Dictionary of Military and Naval Quotations , United States Naval Institute, Annapolis, MD,1966.

[24] Newell, C.R., "Fog and Friction, Challenges to Command and Control", Military Review, August 1987.

[25] Van Creveld, M., Command in War, Harvard University Press, Cambridge, MA, 1985.

[26]U.S. Department of the Army, Field Manual 100-5, Operations, Washington, DC, May 1986.

[27]Watkins, J.D., "The Maritime Strategy", Proceedings of the U.S. Naval Institute, Annapolis, MD, January,1986.

[28] U.S. Department of the Air Force, AFM 1-1, "Functions and Basic Doctrine of the United States Air Force", Washington, DC,
February 1979.

[29] U.S. Department of the Air Force, TACM 2-1,"Tactical Air Operations", Washington, DC, April, 1978.

[30] Douhet, G., The Command of the Air, translated by Dino Ferrari, Coward-McCann, Inc., New York, NY, 1942.

[31] Mitchell, W., Skyways-A Book in Modern Aeronautics, J.B. Lippincott Co., Phila., PA, 1930.

[32] de Seversky, A.P., Victory Through Air Power, Simon and Schuster, New York, NY, 1942.

[33] Mahan, A.T., The Influence of Sea Power Upon History -1600-1783, 1890.

[34] Mahan, A.T., The Influence of Sea Power Upon The French Revolution and Empire, 1793-1812, 1892.

[35] Mahan, A.T., Sea Power and its Relation to the War of 1812, 1905.

[36] Earle, E.M. (ed.), Makers of Modern Strategy: Military Thought from Machiavelli to Hitler, Princeton University Press, Princeton, NJ, 1952.

[37] U.S. Joint Chiefs of Staff, Publications 1,"Dictionary of Military and Associated Terms", Washington, DC, January, 1986.

[38] Lanchester, F.W., "Aircraft in Warfare: The Dawn of the Fourth Arm-No. 5, The Principle of Concentration", Engineering 98, 1914.

[39] Morse, P.M. and Kimball, G.E., Methods of Operations Research, MIT Press, Cambridge, MA, 1951.

[40] Hughes, W.P., Jr., (ed.), Military Modeling, Military Operations Research Society, Alexandria, VA, 1984.

[41] Signori, D., and Starr, S., "The Mission Oriented Approach to NATO C2 Planning", SIGNAL, 1987.

[42] Torelli, P., "A Framework for Assessing DOD C3 Programs", [15]

[43] Griswold, D.B., "Mission Oriented Planning to Support the Maritime Major NATO Communities", [15].

[44] Sovereign, M., and Hingorani, G., "Long-Term NATO C3 Planning", [15].

[45] Sweet, R., Mensh, D., Gandee, P., Stone, I., and Briggs, K., "The Modular Command and Control Evaluation Structure (MCES): Applications of and Expansion to C3 Architectural Evaluation", Naval Postgraduate School, Monterey, CA, September, 1986.

[46] Sweet, R., "The MCES and the Search for Generic Measures", [13].

[47] SAL Sweet, R., and Sovereign, M., "Evaluating Command and Control: A Modular Structure", C3I Handbook, Edition Three, Defense Electronics, 1988.

[48] Sweet, R., and Levis, A., "The Modular Command

and Control Evaluation Structure (MCES) and Command Center Evaluations: Case Study: Proof-of-Concept/Experimental test bed (POC/ET)", 1987 IEEE Military Communications Conference, Washington, DC, October 1987.

[49] Gandee, P.L., Gray, M. D., and Sweet, R., "Evaluating Alternative Air Defense Architectures", SIGNAL, January, 1987.

[50] Bouthonnier, V. and Levis, A.H., "Effectiveness Analysis of C2 Systems", IEEE Trans. on Systems, Man and Cybernetics,
Vol. SMC-14, No. 1, January-February 1984.

[51] Levis, A.H., "System Effectiveness Analysis", Laboratory for Information and Decisions Systems, M.I.T., Cambridge, MA,1987.

[52] Lawson, J.S., "A Unified Theory of Command and Control", 41st Military Operations Research Symposium, 11-13 July,1978.

[53] Lawson, J.S., "The State Variables of a Command Control System", [3].

[54] Lawson, J.W., "The State of Variables of a Command Control System", Analytical Concepts in Command and Control, MORS Monograph, 1981.

[55] Wohl, J.G., "Force Management Decision Requirements for Air Force Tactical Command and Control", IEEE Transactions on Systems, Man, and Cybernetics, Vol. SMC-11, No. 9, Sept. 1981.

[56] Wohl, J.G., Gootkind, D. and D'Angelo, H., "Measures of Merit for Command and Control", MITRE Report MTR-8217, Bedford, MA, 15 January 1981.

[57] Sweet, R., Metersky, M., and Sovereign, M., "Command and Control Evaluation Workshop", Military Operations Research Society, Naval Postgraduate School, Monterey, CA, 1985.

[58] Mayk, I., and Rubin, I., "Paradigms for Understanding C3, Anyone?", [13].

[59] Mayk, I., Rosenstark, S., and Frank, J., "Analysis of C3 Systems Based on a Proposed Canonical Reference Model", [12].

[60] Smith, Charles L. "Survey and Assessment of the Role and Use of Modeling in the Analysis of Command and Control Systems",[13].

[61] Carter, A.B., Steinbriner, J.D., and Zraket, C.A., (eds.) Managing Nuclear Operations, Brookings Institute, Washington, DC, 1987.

[62] Cothier, H. P., and Levis, A.H., "Assessment of Timeliness in Command and Control", [11].

[63] Lawson, J.S., "The Role of Time in a Command Control System", [7].

[64] Rubin, I., Baker, J., and Mayk, I., "Stochastic C3 Modeling and Analysis for Multi-Phase Battle Systems", [13].

[65] Rubin, I., and Mayk, I., "Dynamic Stochastic C3 Models and Their Performance Evaluation", [12].

[66] Rubin, I., and Mayk, I., "Modeling and Analysis of Markovian Multi-force C3 Processes", [12].

[67] Ingber, L., "Nonlinear Nonequilibrium Statistical Mechanics Approach to C3 Systems", [12].

[68] Ingber, L., and Upton, S., "Stochastic Model of Combat", [13].

[69] Ingber, L., "Applications of Biological Intelligence to Command, Control and Communications", in Computer Simulation in Brain Science: Proceedings, University of Copenhagen, 1986.

[70] Ingber, L., "Mathematical Comparison of Computer-Models to Exercise Data: Comparison of JANUS(T) to National Training Center Data", [15].

[71] Goodman, I.R., "A Probabilistic/Possibilistic Approach to Modeling C3 Systems", [12].

[72] Goodman, I.R., "A Probabilistic/Possibilistic Approach to Modeling C3 Systems"; Part II, [13].

[73] Goodman, I.R., "Toward a General Theory of C3 Processes", [15].

[74] Coe, G.Q., and Dockery, J.T., "OJCS Initiatives in C2 Analysis and Simulation", Science of Command and Control, [14].

[75] Dockery, J.T., "A Fuzzy Set Treatment of Effectiveness Measures", TM 729, SHAPE Technical Center, The Hague, Netherlands, January,1984.

[76] Woodcock, A.E.R., and Dockery, J.T., "Application of Catastrophe Theory to the Analysis of Military Behavior", SHAPE Technical Center, STC; CR-56, The Hague, Netherlands, 1984.

[77] Woodcock, A.E.R., and Dockery, J.T., "Mathematical Catastrophes and Military Victories", SIGNAL, April, 1987.

[78] Gleick, J., Chaos, Making a New Science, Viking, Penguin, Inc., New York, NY, 1987.

[79] Strack, C., Elements of C3 Theory, draft report for Defense Communications Agency, 30 January 1985.

[80] Cushman, J.H., "A Field Commander's Views on Measures of Merit", MITRE Corp., Bedford, MA, 4 December 1978.

[81] Serfaty, D., Athans, M., and Tenney, R., "Towards a Theory of Headquarters Effectiveness", [15].

[82] Reisig, W., Petri Nets: An Introduction, Springer Verlag, New York, 1982.

[83] Peterson, J.L., Petri Net Theory and the Modeling of Systems, Prentice-Hall, Englewood Cliffs, NJ, 1981.

[84] Levis, A.H., and Tabak, D., "Petri Net Representation of Decision Models", [10].

[85] Tabak, D. and Levis, A.H., "Petri Net Representation of Decision Models", IEEE Transaction on Systems, Man and Cybernetics, Vol. 5, SMC-15, No. 6.

[86] Levis, A.H., and Hillion, H.P., "Performance Evaluation of Decisionmaking Organizations", [13].

[87] Levis, A.H., and Andreadakis, S.K., "Design Methodology for Command and Control Organizations", [13].

[88] MacMillan, J., and Wohl, J.G., "Human Decisions in an SDI Command and Control System: A Petri Net Model", [15].

[89]ALF Madsen, K.R., "Modeling and Simulation of Army Division Headquarters Using Petri Nets", [15].

[90] Levis, A.H. and Boettcher,K.L., "Decisionmaking Organizations with Acyclical Information Structures,"[8]

[91] Boettcher, K.L.and Levis,A.H., "Modeling the Interacting Decisionmaker with Bounded Rationality", IEEE Transaction Systems, Man and Cybernetics, May/June 1982.

[92] Hall, S.A. and Levis,A.H. "Information Theoretic Models of Memory in Human Decisionmaking Models", [9].

[93] Levis, A.H., "Information Processing and Decisionmaking Organizations: A Mathematical Description", [9].

[94] Kahn, C., Kleinman, D., and Serfaty, D., "Distributed Resource Allocation in a Team", [13].

[95] Taylor, J.G., Lanchester Models of Warfare, Vols. I and II, ORSA Arlington, VA, 1983.

[96] Wozencraft, J.M., and Moose, P.H., "Lanchester's Equations and Game Theory", [9].

[97] Wozencraft, J.M., and Moose, P.H., "Characteristic Trajectories of Generalized Lanchester Equations", NPS 62-87-014, U.S. Naval Postgraduate School, Monterey, CA, June 1987.

[98] Shubik, M., (ed.), Mathematics of Conflict, Elsevier Science Publishers, New York, NY, 1983.

[99] Protopopescu, V., et al, Combat Modeling with Partial Differential Equations, Oak Ridge National

Laboratory Report ORNL/TM 10636, Nov. 1987.

[100] Karr, A.F., "Lanchester Attrition Processes and Theater-Level Combat Models", in Mathematics of Conflict,
Elsevier Science Publishers, New York, NY, 1983.

[101] Tavantzis, J., Rosenstark, S., and Frank, J. "A Stochastic Model of Lanchester's Equations", [11].

[102] Shultz, G. W., and Tsokos, C.P., "Stochastic Modeling-Lanchester's First Law", in Tsokos, C.P., and Thrall, R.M., "Decision Information", Academic Press, New York, NY, 1979.

[103] Dockery, J.T., and Santoro, R.T., "Lanchester Revisited: Progress in Modeling C2 in Combat", *SIGNAL*, July 1988.

[104] Isaacs, R., Differential Games, John Wiley, New York, 1965.

[105]ELS Shubik, M. (ed.) "Game Theory: The Language of Strategy?" in Mathematics of Conflict, Elsevier Science Publishers, New York, NY, 1983.

[106] Davenport, W.B., and Root, W.L., Random Signals and Noise, McGraw-Hill, New York, NY 1958.

[107] Gallager, R.G., Information Theory and Reliable Communication, John Wiley and Sons, New York, NY, 1968.

[108] Wozencraft, J.M., and Jacobs, I.M., Principles of Communication Engineering, J. W. Wiley, New York, NY, 1965.

[109] Van Trees, H.L., Detection, Estimation and Modulation Theory, Vol. I, J. W. Wiley, New York, New York, 1968.

[110] Van Trees, H.L., Detection, Estimation and Modulation Theory, Vol. III, J. W. Wiley, New York, New York 1971.

[111] Helstrom, C.W., Statistical Theory of Signal Detection, Pergamon, London, 1960.

[112] Berkowitz, R.S., Modern Radar: Analysis, Evaluation and System Design, J.W. Wiley, New York, New York, 1965.

[113] Proceedings of the Tri-Service Data Fusion Symposium, Vol. I, John Hopkins University Applied Physics Laboratory, Laurel, MD, 9-11 June 1987.

[114] Shannon, C.E. and Weaver, W., The Mathematical Theory of Communication, The University of Illinois Press, Urbana, Ill.,1949.

[115] Augustine, N.R., Augustine's Laws Viking, New York, NY, 1986.

5

Organization Theory and C³

John P. Crecine and Michael D. Salomone

Introduction

Our approach to command and control assessment draws heavily on the Carnegie School of organization theory as developed in the classic works of Simon, Cyert, and March.

Modern warfare represents the clash of two, rival organizations, each bent on containing or destroying the other. Adopting an organizational perspective for evaluating competing military organizations—such as the NATO and Warsaw Pact organizations—generates a different set of questions and focuses attention on different issues than the traditional forms of force balance calculations and threat assessment. In particular, the focus shifts from relative firepower, technological properties of weapons systems, and numbers of men and equipment to the organizational capabilities of military units and to the ways in which the various components of military units act in a coordinated way.

What can the complex of men and equipment do, how well, how quickly, and under what conditions? A central issue in determining the organizational capabilities of a large, military force concerns the aggregate division of labor in the force: What are the component units and what are their special capabilities? For a given division of labor, how do the component units work together and how are they coordinated so as to perform a set of functions beyond the capabilities of any single unit? [1]

A second set of issues is the natural by-product of organization structure; namely, formal organizational divisions or "seams." Formal organizational boundaries are important in that they usually define areas where all aspects of a subunit's operations, "within boundaries," are organic to the subunit, whereas operations involving other subunits—"across boundaries"—involve explicit inter-unit coordination and planning. Organizational boundaries—the "seams" in the organization—are important to a military offense because they define what units require explicit coordination from a command point of view. Organizational seams are important to a military organization in a defensive posture because they represent points where the disruption of explicit, inter-unit communications in real time will do more to disrupt overall capabilities than the disruption of intra-unit communications. This is simply because inter-unit coordination more often involves explicit communication and coordination, and intra-unit coordination more often involves tacit forms of coordination in addition to, but often in place of, explicit communication.

Coordination is the Achilles' heel of the organizational strategy of overcoming individual cognitive limits through specialization and division of labor. The efforts of numerous individuals and separate groups of individuals must be structured to form coherent and useful patterns of activity. But this structuring task itself can easily become so complex as to overwhelm the cognitive capacities of those who must carry it out. Indeed, the high frequency of "coordination failures" suggests that such problems commonly occur.

The military command and control process embraces all of the information processing activities associated with monitoring the environment for problems and opportunities, formulating alternatives, deciding between alternatives, issuing mission orders and instructions, and monitoring and controlling the execution of assigned missions. All of these tasks involve coordination among many parts of the command organization, both vertically within the chain of command, and horizontally beyond it into the chain of command of equivalent or parallel organizations. Adjacent army groups, corps, and divisions are examples of this latter point. The additional coordination requirement for air forces to support ground forces through battlefield air interdiction or close air support missions adds another layer of complexity. Air forces within NATO not only have their own command structure, but they operate with radically different equipment, training, doctrine, requirements, and ethos than do ground forces.

In adopting an organizational focus for C3I research, attention is immediately drawn to the question of what is being coordinated with what and the prominent role that the formal structure of the organization plays. An organizational approach to C3I is particularly important in two different types of situations: 1) where the functioning of an organization depends crucially upon explicit coordination, communication, command and control, across significant organizational boundaries; and 2) where the functioning of a subunit of the organization is automatic or involves the execution of standard operating procedures.[2]

The Organization of NATO in Central Europe

The current organizational structure of NATO evolved from political agreements about the postwar structure of Europe, the formation of the Atlantic Alliance, and the sensitivities of individual member nations. While it has been relatively stable concerning national assignments within the command structure of the NATO military force, the NATO military structure also reflects the growing importance and military strength of the Federal Republic of Germany and the worldwide economic and military strength of the United States. It is anything but conducive to the integration of the multinational forces into a single, well-coordinated military force.

SACEUR (Supreme Allied Commander Europe, a U.S. general officer who is also the U.S. Commander in Chief, Europe, in peacetime) is the most prominent of three distinct major commanders in NATO. The Supreme Allied Commander Atlantic (SACLANT) and Commander in Chief Channel (CINCHAN) are the other two major commands. There are six separate commands on the level beneath SACEUR in the NATO hierarchy, each of these

with two or three separate commands beneath it, and each filled by a military commander of a different nationality.

In the NATO Central Region—which includes the geographic territory of Belgium, the Federal Republic of Germany, Luxembourg, and the Netherlands—the joint forces commander is Commander-in-Chief Allied Forces Central Europe (CINCCENT). He is a German general officer, one level below SACEUR in the NATO structure. His command is further subdivided into: the Northern Army Group (NORTHAG, commanded by a British general officer who is also Commander, British Army of the Rhine, in peacetime), Allied Air Forces Central Europe (AAFCE, with a U.S. commander who is also commander of U.S. Air Force, Europe in peacetime), and the Central Army Group (CENTAG, also commanded by a U.S. general officer who serves as CINC U.S. Army, Europe, in peacetime). AAFCE has within it the separate commands of Second and Fourth Allied Tactical Air Forces (TWO and FOURATAF). TWOATAF, to a great extent, reflects the operating style of the Royal Air Force although its headquarters organization reflects the multinational character of the forces (Belgium, Netherlands, German, U.S., and U.K.) it will command in wartime. FOURATAF, on the other hand, is dominated by the Unites States Air Force and reflects U.S. organizational and operating style, attitudes, and concepts of operations.

This highly disintegrated chain of command is largely a product of, and is further exacerbated by, political considerations. There is an absolute requirement within NATO, institutionalized through political reality and administrative procedure, to preserve a degree of national sovereignty in military affairs, a constraint that does not impede the Warsaw Treaty Organization. An additional, potentially problematic characteristic of the NATO deployment of forces in Central Europe is the division of the East-West border into territorial slices, "layer cake" fashion, with each layer defended by the forces of one nation. Thus, at minimum, the forces of five nations (the U.S., West Germany, the Netherlands, the United Kingdom, and Belgium) will be arrayed along the front. It is possible that if the NATO position began to deteriorate, the French would also move forward to take a slice of their own, or be employed as a reserve coming in on top of any of the national corps sectors, or at a corps boundary.

Even from this superficial description of the organizational structure at NATO echelons above corp, it is simple to conclude that the C[3] and coordination requirements are enormous for fighting a joint and combined conventional war with a multinational alliance in Central Europe. That these requirements be met is also of the utmost importance as more and more emphasis is placed upon the importance of conventional defense of Europe to support deterrence. C[3] failures make cheap victories possible, as the fall of France in 1940 and the collapse of South Vietnam's armed forces in 1975 make clear.

The Carnegie School of Organization Theory

The analysis of command and control presented here is based on the tenets of "organization theory" derived from an analysis of the behavioral limits and capabilities of individuals and of organizations. Drawn from the classic writings of Richard Cyert, Herbert Simon, and James March, the "Carnegie School" of organization theory is a set of propositions, inferences, and hypotheses drawn from research in cognitive psychology, information processing, and human behavior and adapted to the description, analysis, and measurement of behavioral variance in collections of individuals commonly known as organizations. This approach, well known in its application to problems of business and industrial management, has been applied by us to problems of command and control, most recently focusing on the organizational seams in the NATO Central Region from theater to division level.

Bounded Rationality

The Carnegie approach rests on a key premise: Certain fundamental cognitive constraints severely limit the human capacity for rational action. Limits on the amount of information that can be attended to force individuals to respond to only limited aspects of their environment.[3] And limits on the ability actively to manipulate conceptual information force individuals to rely on relatively simple mental strategies that frequently violate conventional notions of rational inference and choice.[4]

Organizations may be viewed as devices for overcoming individual cognitive limits. Through specialization, individuals acquire the capacity to apply relatively complex cognitive strategies to narrowly defined task environments.[5] Through division of labor, substantial cognitive resources can be simultaneously brought to bear on many tasks or information sources at a time.[6]

Central to this perspective on organizations is the concept of "bounded rationality"; that is, the idea that the limits of human capacity to generate alternatives, process information and solve problems constrains the organizational decision making process. The five summary points which follow attempt to capture the direction of Simon's thinking on this concept.

First, because most problems that an organization faces are complex, they are factored, or split into quasi-independent parts and dealt with individually, usually by separate units of an organization.

Second, optimization, or finding the best alternative, is replaced by "satisficing." This means choosing the first alternative which meets minimum decision criteria, that is, acting on the first alternative that is "good enough." One does not look for every needle in the haystack, only for the first one that is sharp enough.

Third, because satisficing requires stopping at the first alternative that is good enough, the order in which organizations develop and propose alternative solutions to problems is a critical determinant of which alternative will be chosen.

Fourth, organizations seek to avoid uncertainty, particularly uncertain future consequences of present actions. Thus, decisions which maximize short run feedback are preferred.

Fifth, repertories of responses or routines generally define the range of organizational choice in recurring or routine situations. Thus, the organization will search for analogs of the familiar when confronted with the unfamiliar.

Concentrating on the bounded character of human rationality, Cyert, March, and Simon focus their attention on the consequences for the decision environment of less than complete information, limited information processing capabilities, uncertainty avoidance, sequential search procedures, adherence to routine, and satisficing as an alternative selection mechanism. Thus, organizational structure and conventional practice become critical factors in the development of goals, the formulation of expectations, and the execution of choice.[7]

An organization can be viewed as a complex system of parts each with specialized capabilities. When an organization is functioning in a coherent fashion, it is acting according to an implicit or explicit plan or strategy for coordinating the behaviors of the various organizational parts. The overall division of labor and coordination strategy for an organization is reflected in the structure of the organization. For an organization, its structure and division of labor defines what must be coordinated with what. Once the boundaries between various subunits are identified and the functional interdependencies—division of labor between the various subunits—are known, it becomes clear where the key coordination points are.

The Coordination Problem

The potential accomplishments of human organizations are almost limitless. In practice, however, human organizations must overcome a fundamental difficulty—namely, the necessity of assuring that the pattern of activities being carried out by individuals in various subunits of the organization fit together in a relatively coherent fashion that results in progress toward the organization's fundamental objectives. Achieving such coherence defines what is termed the coordination problem. But this structuring task itself can easily become so complex as to overwhelm the cognitive capacities of those who must carry it out.[8] Indeed, the high frequency of "coordination failures" suggests that such problems commonly occur.

Coordination Strategies

There are four basic mechanisms for achieving coordination in human organizations.[9] *Direct supervision* is the most obvious. Here, a supervisor exercises a degree of control over the behavior of subunit members by allocating resources to subunits and by directly influencing certain details of the subordinates' behavior. Using the *mutual adjustment* method, two or more actors agree to share resources and to confer with one another concerning decisions that affect the activities of those involved. *Standardization,* or the creation of standard operating procedures (SOPs), offers a third mechanism for achieving coordination.[10] A supervisor can exercise control over the behavior of subunit members by creating standard procedures governing the behavior of subunit members under specified conditions. Thus, the supervisor is able to influence behavior even in situations of which he is unaware. Moreover, subunits can coordinate with one another purely on the basis of shared expectations. No direct communication is necessary if each subunit can anticipate which SOPs other subunits will implement and what the outcomes will be. Explicit *planning* is another way to achieve coordination. It is equivalent to direct supervision in advance. Achieving coordination via resource allocation processes or the creation of SOPs requires significant lead times to accommodate the planning process. Direct supervision and mutual adjustment rely on real-time coordination, which is far more difficult and costly than pre-planning.

Whatever the means chosen, achieving coordination is costly.[11] Human and physical resources are directly consumed by the activities of planning, monitoring, and communicating. Also, different mechanisms for achieving coordination produce unintended side effects that are costly to the organization. Direct supervision processes, for instance, place a heavy information processing burden on those in supervisory roles, and as a consequence, information bottlenecks and delays are likely. Also, because supervisors cannot possibly access all of the information available to their subordinates, this mode of achieving coordination will generally fail to exploit detailed information concerning local circumstances.[12] Mutual adjustment processes, if carried to extremes, can result in extraordinarily high rates of information transmission with every actor communicating with every other actor. The resulting state of information overload can cause delays or total paralysis of the organization. Finally, coordination achieved via standing procedures can result in excessively rigid behavior, particularly in the face of circumstances not anticipated when the SOPs were formulated.

Strategies for Limiting Coordination Costs

In light of the coordination costs mentioned above, organizations invariably devise strategies for limiting the degree of coordination attempted.[13] The strategies they may employ include the following:

•*Reliance on nearly decomposable task structures.* A task structure is strictly decomposable if each subtask can be carried out without regard for how other subtasks are performed.[14] A task structure is nearly decomposable if, in the short run, subtasks can be performed without regard for how others are being performed, and in the long run, depend only on a few aggregate characteristics of how other tasks are being performed.[15] While a degree of monitoring, replanning, and adjustment is required in nearly decomposable task structures, the amount of information that must be gathered and transmitted is much smaller than in highly interdependent task structures.

•*Ignoring interdependencies.* When interdependencies do arise, one crude strategy for reducing coordination costs is simply to ignore the interdependence.[16]

•*Ignoring interdependencies except in extreme cases.* A slightly less extreme strategy for limiting coordination costs is to ignore interdependencies in all but extreme cases.[17] A supervisor might monitor for extreme cases of negative interaction between subunits, or the subordinate units themselves might agree to inform one another when their behavior falls outside of agreed upon norms. In a standing procedure mode, SOPs themselves might specify conditions under which supervisors and subunits will transmit information concerning circumstances that fall outside the bounds of the anticipated or acceptable.

•*Creating buffer stocks of slack resources.* Many coordination problems arise when one subunit requests resources or assistance that are to be provided by another subunit. Absent detailed coordination between these subunits, there is a distinct likelihood that short-term shortages will result in delays in filling such requests. One crude solution to this problem is to hold large buffer stocks of slack resources to be used for meeting such demands.[18]

•*Reliance on flexible, general purpose resources within subunits.* To the extent that subunits possess flexible, general purpose resources, they are less likely to require assistance from other subunits, and are less likely to need to coordinate with them.[19]

•*Reliance on standardization as the least cost means for achieving coordination.* Implementation of SOPs is a very inexpensive means for achieving coordination. To the extent that behavior is coordinated via mutual expectations rather than actual communication, coordination can be almost cost-free, apart from the loss of flexibility inherent in standardization.[20]

Whatever the merits of the above strategies for limiting coordination costs, they are not without costs of their own. The planning processes involved in creating a decomposable task structure are not cost-free. The costs of ignoring interdependencies between subunit activities are obvious. At the extreme, ignoring interdependence may result in situations in which subunits carry out their tasks in manners that are directly at cross purposes. For example, the tragic Apollo 1 fire that killed three astronauts was facilitated by coordination failures between design groups. One chose a 100 percent oxygen environment for the capsule while another chose to use materials that were inflammable in most environments, but highly flammable, virtually explosive, in fact, in the 100 percent oxygen atmosphere of the capsule.[21]

The costs of ignoring all but exceptional cases of interdependence are similar to those of ignoring interdependence altogether, though if the exceptional cases are well chosen, the resulting costs may be considerably smaller in magnitude. Maintaining stocks of slack resources is costly in that the human and capital resources held in reserve are essentially wasted until they are utilized, and there may be inventory costs associated with holding capital resources. Relying on flexible, general purpose resources also is costly because such resources are likely to be more expensive to procure and maintain than specialized resources [22] and, because of their complexity, are likely to suffer from poor reliability. The recent difficulties associated with procuring multipurpose fighter aircraft are instructive in this regard.[23]

Finally, despite the advantages of relying on standing procedures, this strategy for simplifying coordination processes may result in excessively rigid behavior, especially in the face of an uncertain and rapidly changing environment.[24] Attempts to overcome this shortcoming by devising finely-tuned SOPs, sensitive to changing environmental conditions, will result in substantial planning costs. And if the resulting SOPs are too complex, they are likely to be poorly implemented. The basic message here is to keep it simple. This strategy, however, runs the risk of not planning for a sufficient range of contingencies. Once again, the key in organizational design is to strike a balance between competing goods.

The Virtues of Hierarchy

Designers of organizational decision and control systems are faced with a difficult trade-off between coordination costs and the costs of attempting to limit coordination.[25] Almost invariably, organizations resolve this design dilemma by developing hierarchical task and control structures.[26] Hierarchical control systems are particularly advantageous when the task structure is decomposable, or nearly so. Hierarchical control systems require less communication than other types of control systems. If communication is strictly hierarchical, each actor communicates with only one superior and a small number of subordinates. Furthermore, the complexity of a hierarchical organization is nearly constant across all levels as far as the individual actor is concerned, which is advantageous since the ability to think about more than one complex task at once is not likely to vary significantly among individuals at different levels in the hierarchy. Nearly decomposable hierarchical systems, composed of stable subsystems, have an inherent evolutionary advantage since a change in one component will require little or no adjustment in the others, and damage to one component can be repaired without any alteration of the others. In short, hierarchy substantially simplifies planning and control processes, and it promotes survival in the face of a changing or hostile environment. Whatever the balance struck between the costs of coordination and the costs of limiting coordination, the optimal solution is likely to possess many attributes of a nearly decomposable hierarchy.

Centralization and Decentralization: Trade-offs and Mixed Organizational Strategies

Discussions of what constitutes a good decision structure or process also frequently revolve around the issue of centralization of control. The degree of centralization is conceptually defined in terms of the number of alternatives available to subordinates that satisfy the basic resource constraints imposed by the superior, and that are compatible with the goals and instructions established by the superior.[27] The greater the number of alternatives, the less the degree of centralization.

Most control systems are centralized in some respects and decentralized in others. The U.S. Army's mission planning procedures, for example, involve centralized specification of mission objectives and resources, but decentralized planning and execution of mission details. The degree of centralization may also vary by task. Nuclear weapons maintained by the U.S. Army operate under much tighter control than do conventional weapons under Army control.

Higher level control may be of two sorts.[28] The higher authority may physically constrain the options available to subordinates by means of resource allocation decisions. Or, the higher authority may influence the actual decisions of subordinates by transmitting to them goals, task instructions, or information about the internal and external environment of the organization. If effective, the second approach is likely to be more flexible and precise than the first. Invariably, the actual method of control will involve some combination of the two.

The primary advantage of centralization is the increased potential for coordinating subunit activities. The liabilities of direct supervision include delay and the failure to fully exploit local information and expertise. The centralization issue can also be framed in terms of organization based on mission versus organization based on function.

Function versus Mission Decomposition

An organizational hierarchy can be structured in many ways. One critical choice is whether to decompose by function or by mission. Functional decomposition is decomposition in terms of area of specialization. Immediately beneath the top level commander are specialties; then at the next level, subspecialties; and so forth. The U.S. Army can be thought of as being structured in terms of major specialty areas such as infantry, armor, artillery, etc. These specialties can be further divided into subspecialties; for instance, airborne infantry, air mobile infantry, mechanized infantry, and light infantry.

Mission decomposition is decomposition in terms of task or purpose. Thus, the overall commander supervises a group of major mission commanders who in turn oversee a set of submission commanders, and on down the line. The armed forces of the U.S. can be thought of in terms of a number of missions: strategic nuclear deterrence and retaliation; continental air defense of the U.S.; and defense of Europe against Soviet attack, to name a few. Each of these major missions can be repeatedly subdivided, of course, in terms of increasingly specific tasks.

In practice, most organizations use a mixture of organization by mission and organization by function. The most common pattern is to organize by mission at the top of the hierarchy and by function nearer the bottom. In any organization that employs a mixture of mission and functional decomposition, the central design question is at what level the decomposition should proceed in terms of missions and at what level according to specialty and function. A second critical question is how the specialty subunits can be brought into the service of the mission-oriented subunits. There is, of course, no unique answer to either question. Differences of place and circumstance dominate.

The primary advantage of functional decomposition is that it permits the organization to exploit the economies of scale inherent in high degrees of specialization. The major liability of functionally-organized units is that there are very few missions that they can perform by themselves. To perform a novel mission, using functionally-organized subunits, a higher level decision maker must provide overall coordination and control between the units involved in the process, or SOPs may substitute if the task is a routine one.[29] Organizations that are functionally structured from top to bottom are likely to incur heavy coordination costs whatever the means employed to achieve that coordination. Because of the need for coordination, functionally structured organizations tend toward inflexibility and inability to adapt to rapidly changing situations.

The primary alternative to functional decomposition is to organize by mission at the top, and by function nearer the bottom of the hierarchy. An obvious cost of this approach is that it reduces the opportunity for exploiting economies of scale, since an organization cannot afford to include highly trained specialists or exotic technology in every major operational subunit.[30] Mission decomposition has other potential drawbacks as well. Mission-oriented subgroups may become parochial in outlook, overestimating the value of their mission relative to other organizational missions.[31] Finally, to the extent that mission-based decomposition results in a highly decentralized command structure, the resulting ambiguity and requirements for unstructured problem-solving by lower level personnel may result in degraded morale or performance if the task requirements exceed the capabilities of the individuals involved. But mission decomposition can substantially reduce coordination requirements, provided that the set of missions involved is decomposable or nearly so. Higher-level decision makers need only to assign tasks and allocate resources. Details of execution can be left to the discretion of mission and submission commands, permitting maximum exploitation of local information. It is clear, then, that notions of centralization and decentralization are closely linked to whether an organization is structured in terms of mission or function. Mission-based decomposition encourages decentralization, provided that the missions are relatively independent. Function-based decomposition requires higher degrees of coordination between subunits, and hence tends to result in a highly centralized command structure.

Organizational Seams

The various coordination strategies discussed in the preceding section are meant to coordinate organizations, or parts of organizations, despite structural, functional, institutional, or national divisions between them. These "seams" between organizations can be either minor or major impediments to coordination attempts. Coordination within one unit of an organization is "organic," is often rehearsed, often involves tacit rather than explicit coordination procedures, and is generally easier to accomplish. Individuals or groups within a subunit know each other, know what to expect from each other's behavior, often can anticipate behavior, and often rely on implicit coordination. Coordination across subunit boundaries is often done explicitly and generally involves real coordination costs or detailed plans. Coordination across organizational boundaries tends to be rehearsed less frequently than coordinated activities within subunit boundaries and is generally much more difficult for an organization to accomplish.

Assuming that different subunits of an organization must act together, coordinating the activities of heterogeneous units—for example, a ground unit and an air wing—is more difficult than coordinating two similar units—two adjacent U.S. Army infantry battalions, for example. Coordination between similar units is inherently easier because each unit has a "conceptual model" for the behavior of their counterparts, in essence, "They behave just like we would."

Illustrative of this approach is the provision of fire support, which can be provided from the ground by artillery that is organic to the ground forces, or from the air by air forces which exist in the U.S. armed forces as a different branch of service. The differences in artillery versus air support are representative of the differences between major and minor organizational seams. Artillery is indigenous to the ground forces. Artillery officers have the same basic and doctrinal training as the forces they support. Their command structure is completely integrated with that of the ground commanders, and their communications equipment can easily access the main communications network. Tactical air support, in contrast, is provided by an external organization, namely the Air Force. Air Force officers come from a different system of military education and background, and their attitudes and preferences reflect different incentives and belief systems. The command structures and planning cycles differ radically between the two organizations, and they have divergent views on allocation of resources among missions and acquisition of types of airframes and ordnance to support these mission priorities.

The difficulties in *coordinating the air and ground wars* further illustrate the coordination problem within NATO. The Air Force and the Army take fundamentally different positions on what form air support for ground forces should take. The Air Force believes that it cannot effectively support battlefield air interdiction and close air support along the forward edge of the battle area until it has established air superiority. The Army wants to see more direct applications of air power in support of the troops. In addition to these philosophical differences, there are methodological ones. The Army and Air Force operate on entirely different planning cycles with different time lines. The differences in time lines and planning horizons can have a significant and adverse effect on the interactions between the two services. For example, the ground commander's plans may be severely disrupted or even cancelled if his specific requests for air support or reconnaissance cannot be fulfilled within the time frame required.

Seams in an organization may be either "horizontal" or "vertical" in orientation. The seams between two units on the same level of a hierarchy are considered to be horizontal; seams between actors at different levels are considered vertical. The "height" of the seams, measured as the degree of difficulty in overcoming them, may also vary. The coordination of two infantry battalions in the same regiment, for instance, may be substantially easier than the coordination of two battalions in different divisions. The most insurmountable seams, perhaps, appear across multinational boundaries. To illustrate this point, note that British aviators in TWOATAF operate under different assumptions and doctrine than their United States counterparts in FOURATAF. British air forces assume that during wartime, they will be operating for some time when they are out of contact with higher command levels. Their training reflects this assumption. In contrast, U.S. procedures assume a high degree of contact with flight controllers throughout a mission and that squadron coordination will be accomplished through real-time communication links.

In addition, the different national military

organizations in NATO have nationally provided and therefore different intelligence and logistical support systems on which they will rely during wartime, while operating under the NATO command. These national forces use different communications systems and weaponry, and have different levels of readiness. It is clear that the seams between the elements of the NATO multinational force are inherently "higher" than the seams that separate different units of the Bundeswehr.

Coordinating the Offense versus Coordinating the Defense

The kind of coordination required generally varies with the task or function that the organization is trying to perform. For example, in coordinating the activities of various military units in an offensive operation, coordination itself is a command task. In a wartime situation involving two large, competing military organizations, the environment is likely to be uncertain and rapidly changing. Coordinating military organizations having highly specialized functions and equipment in a complex, hostile environment is inherently easier if achieved through preplanning than if the complex of deployment, logistical, strategic and tactical decisions, mission assignments, and force-tasking has to be done in real time, "on the fly." Real-time coordination requires accurate real-time intelligence, assessment of a fluid battlefield environment, reliable and timely communications, and competent and continuous exercise of command.

The responsible military commander generally has a plan, reflected in the orders given to the various components of the organization. If the military organization on the offense is able to control the pace of battle, its coordination is accomplished simply by following the direction of the overall military commander or by following the offensive plan. Coordination is generally done through preplanning or the transmission of simple, easily understood orders.

In contrast, coordination of military units involved in defensive operations generally cannot rely on elaborate preplanning in order to accomplish their task, unless they can predict with sufficient confidence, the exact location, size, and timing of the enemy attack. Explicit forms of communication are generally required, and if the communications between various units can be disrupted, delayed, or if the link can be destroyed, the capability of the defensive military organization to operate in a coordinated capacity is severely diminished, if not eliminated. Coordination by explicit communication plays a much larger role in a defensive organization than an offensive one because defensive military operations are reactive, by definition.

The essence of Soviet military doctrine is to control the pace of battle through offensive operations that rely on preplanning as the primary coordination strategy. By creating an environment that the defense must adapt and react to, the Soviets hope to compound the coordination problem for their enemy. By keeping the pace of their offense rapid, the Soviets seem to be trying to force the defense to rely exclusively on real-time coordination strategies that are inherently more costly and difficult.

Six Key Attributes of Effective Command and Control

In previous research on command and control in the NATO Central Region— sponsored by OSD/Net Assessment—Coulam, Crecine, Fischer, and Salomone defined six essential properties of a command and control system.[32] The list of essential properties could certainly be expanded, but research has indicated that these six characteristics bound the minimum requirements for effective command and control. The properties were placed in two groups. The first group consisted of "performance characteristics" and included 1) processing capacity, 2) reaction time, 3) flexibility, and 4) coordination capability. The second group included the "continuity of command" characteristics of 5) physical survivability and 6) robustness against attrition.

With respect to *processing capacity*, research shows that a C^3 system should have as much cognitive, computational, and communications capacity as possible. This requirement is affected not only by numbers of men and levels of equipment and technological sophistication, but also by levels of training and rehearsal, communications discipline, well-developed routines and standard operating procedures.

Noting that organizations cannot do well what they have not rehearsed, the procedural dimension of this function becomes at least as important as the technical. For example, a courier on a motorcycle or in a jeep may be a much more effective communications device than a high speed communications link which can transmit great volumes of information to corps headquarters at such speed that the printers and displays that pull it off the wire there are backed up for twelve hours with incoming messages.

The second critical performance attribute identified was *reaction time*, defined as the amount of time that elapses between the onset of a change in the environment (stimulus) and the initiation of a response. Reaction time is a function of the level of facility gained through rehearsal and procedure, the degree of processing capacity compared to the amount of processing load, and centralization or decentralization of the organization with respect to both structure and decision locus. The greater the ratio of processing capacity to processing load, the faster the reaction time. Perhaps less obviously, the more decentralized a C^2 system, the greater its ability to respond rapidly to changes in local conditions that affect only one or a few subunits.[33]

The third property is *flexibility*, that is, the capacity to adapt to rapidly changing conditions. Like the two performance attributes discussed above, a central key to flexibility lies in training and rehearsal. According to Crecine, et al.:

Flexibility is enhanced by the development and rehearsal of both tactical and C^3 SOPs that cover a broad range of scenarios. This speculation rests on the assumption that the effectiveness of SOPs is inherently scenario dependent. Confronted by unforeseen contingencies, the command system will be forced either to engage in time consuming improvisation or to rely on SOPs that are inappropriate to the conditions of battle.[34]

Coordination capacity is the fourth critical performance characteristic of effective command and control identified by Crecine, Coulam, Fischer, and Salomone. Coordination capacity is determined by a mix of hierarchical and structural variables and procedures. The authors note, for example, that in the field, "C^3 systems differ in their ability to plan and conduct 'combined arms' operations in which dif-ferent types of forces — infantry, tanks, artillery, aircraft — work in conjunction with one another. At a higher level, for instance the whole European Theater, different C^3 systems will differ in their ability to coordinate battles in different sectors, or to reallocate forces and supplies from one major sector to another."[35]

The two continuity of command characteristics identified—*survivability* and *robustness against attrition*—were found to be influenced and enhanced not only by physical means, such as hardening, but also by procedural actions, such as communications disciplines that minimize the use of radios that easily can be detected by enemy forces. Again procedures, SOPs and organizational capabilities enhanced by realistic rehearsal can be effective substitutes for direct real-time communication if the objective is to avoid detection and possible destruction. Tacit communication and shared expectations, perceptions and beliefs can enhance survivability under conditions where intense hostile action is directed at first locating and then destroying headquarters and C^3 systems. Mobility, the separation of command posts from transmission facilities, masking, hiding of facilities, and deception are other means to enhance survivability. But no matter how great the efforts to enhance C^3 survivability, a concerted attack on a C^3 system is bound to be partially successful. Some communications links will be jammed and some components of the system will be destroyed. Any truly survivable system must be able to continue to perform critical functions despite damage or attrition of C^3 system components. Thus, the importance of the second property of continuity of command, robustness, is mainly a matter of organizational structure and procedures.[36]

Redundancy in human and physical C^3 assets, and procedures for devolution of command, are essential if a damaged or degraded C^3 system is to reconstitute itself or degrade gracefully. Since C^3 failures make cheap victories possible, under conditions of massive damage to a command and control system, the "system must be robust in the sense of being able to reconstitute itself to form an effective fighting force from those C^3 elements and fighting units that have survived."[37] This capability will be enhanced by structures and procedures that permit surviving units to fight on their own while the command structure reconstitutes itself. The Crecine, et al. study concludes that the many layers of command that exist in the NATO command and its constituent national corps and below "create a substantial potential for reconstituting at many levels, if the survivors can operate alone in the short run."[38] This need to be able to function independently for short periods of time mandates a nearly decomposable task structure that allows subunits to operate constructively on their own. Robustness is also enhanced by realistic exercise of operations under conditions of degraded communications and the physical loss of headquarters or other command facilities. It is in realistic training that NATO may have one of its greatest inadequacies.

Crecine, et al. concluded that although there are measurable performance attributes associated with the functioning of a command and control system along the six dimensions listed above, the effectiveness of a system of command and control can only be evaluated in the broader context of a specific scenario. The interplay of the battle scenario and the operational concepts employed by each side will be a prime determinant of the appropriateness and effectiveness of the respective command and control concepts, systems, and capabilities of the combatants. For example, with respect to opposing military forces, they conclude that: "...the capacity for centralized coordination is less important to a military force that relies on independent action undertaken by isolated defensive units than to one that plans to establish and hold an unbroken line of defensive forces."[40] This tenet concerning scenario, operational concept and C[2] structure and process is a condition of the interaction of organization and task environment. It is not only an aspect of military organization and command and control, but is germane to any organization and task environment.

Two themes clearly emerged from this research. The first is that actual rehearsal under realistic conditions provides the surest basis for effective action. Rehearsal serves to enhance the degree to which coordination can be achieved through reliance on shared knowledge and expectations. Simplicity is the second hallmark of effective action. Alternatives that create minimal coordination requirements are most likely to succeed. To the extent that direct means of coordination are essential, they should be based on coordination SOPs that are themselves simple and well practiced.

Gross Differences Between NATO and the WTO for Conventional War in the Central Region

When one examines the composition of combat forces making up NATO and the WTO, it is clear that the NATO command structure reflects the multinational composition of the Alliance and the fighting forces assembled. In contrast, Soviet military organization dominates all planning for combat missions and combat forces in the WTO. NATO will have a minimum of five nationalities holding sections of the front line and the Central Region; the Americans, West Germans, British, Belgians, and Dutch.

The multinational character of NATO combat forces creates in itself some rather obvious command, communication, and coordination problems that are different from those faced by an all-Soviet WTO combat force such as the Group of Soviet Forces in Germany. Although there is a degree of cultural and linguistic diversity within the Soviet military, this is a fundamentally different problem for the Soviets than, for example, the coordination of adjacent U.S. and West German forces or U.K. and Belgian forces. The Soviet forces are dominated and unified by the Russian language. Even among non-Russian Soviet officers, the language of their training is Russian.

Another gross difference apparent between NATO and the WTO is the degree of centralization in the command structure. The WTO tends to be more centralized than NATO, with key strategic and tactical decisions being made one or two levels in the hierarchy above that of NATO. In addition, it is clear that the NATO command structure places far greater reliance on real-time communication as a way of coordinating the behavior of the various units making up the overall NATO force. In contrast, the Soviet/WTO organization tends to rely much more on detailed preplanning as a way of coordinating various units. This reflects partly the relative capabilities of the two sides, partly cultural as well as socio/political preferences, and partly different strategic concepts and approaches to conventional warfare. NATO quite clearly sees itself primarily as fighting a defensive war and finding itself in situations where elaborate preplanning is of marginal utility because of the reactionary nature of the defense.

Another clear difference between NATO and WTO command structures lies in the way in which the air/land battle is coordinated. In NATO, the coordination takes place through parallel command structures, one for the allied air forces, and one for the ground forces. The primary level of integration for air and ground commands is at the army group level. In the WTO, both air units and ground support units are organic to the ground forces and utilize the same command structure. Soviet ground commanders at the division and army level each "own" appropriate air forces, whereas their counterparts in NATO may directly command ground forces but the associated air forces are "owned" by another command structure and necessary air support must be requested from the parallel structure.

In general, the issues and problems identified by an organization theory approach to the assessment of command and control in the NATO Central Region will become serious if NATO and the Warsaw Pact are actually faced with fighting a conventional war there. In peacetime, the visibility of these problems and the functions which must occur for effective coordination and command, and purposeful and organized activity to ensure military effectiveness is obscured The use of the organization theory approach for analyzing C^3 functions and identifying C^3 problems effectively demonstrates that these problems are primarily structural and organizational, not technical in nature. Unfortunately, structural and organizational issues are among the most difficult to deal with in peacetime.

Notes

[1]Research that made this paper possible was sponsored by Andrew Marshall, Director of Net Assessment, Office of the Secretary of Defense; and Stuart Johnson, Director, Strategic Concepts Development Center, National Defense University. The authors are grateful to them for their support and encouragement. The authors also wish to thank research assistant Lisa MacAnany for her assistance in the preparation of this manuscript.

[2]See Crecine, J. P. and Salomone, M.D., *AFCENT Command and Control Assessment;* Volume I; *Coordination Across Organizational Seams, Theory and Methodology,* Pittsburgh, PA, Joint Management Services, JMS-TR-89/1, January 1989; and Salomone, M.D. and Crecine, J. P., Volume II,; *Coordination and Operational Effectiveness in the NATO Central Region,* Pittsburgh, PA, Joint Management Services, JMS-TR-89/3, March 1989. This research was sponsored by OSD/Net Assessment and the Command and Control Research Program, Strategic Concepts Development Center, National Defense University.

[3]Norman, D.A., *Memory and Attention,* 2nd edition, New York: John Wiley, 1976.

[4]Simon, H. A., *The Sciences of the Artificial,* 2nd edition, Cambridge: The MIT Press, 1981; Kahneman, D., Slovic, P., and Tversky, A., *Judgement Under Uncertainty: Heuristics and Biases,* Cambridge: Cambridge University Press, 1982; Tversky, A. and Kahneman, D., "The Framing of Decisions and the Psychology of Choice," *Science,* No. 211, 1981.

[5]Chase, W. G. and Simon, H. A., "Perceptions in Chess," *Cognitive Psychology,* 1973.

[6]March, J. G. and Simon, H. A., *Organizations,* New York: John Wiley, 1958.

[7]See Cyert, R. M. and March, J. G., *A Behavioral Theory of the Firm,* Englewood Cliffs, NJ: Prentice-Hall, 1963, Simon, H. A., *Models of Man,* New York: John Wiley, 1957. For an example of this schema applied to the problems of the president as chief executive officer in crises, see Paul Y. Hammond and Michael D. Salomone, *Constraints and Boundary Conditions Affecting C^3I Survivability,* Marina del Rey: The Analytical Assessments Corporation, 1981.

[8]March and Simon, *Organizations,* 1958.

[9]Mintzberg, H., *The Structure of Organizations,* Englewood Cliffs, NJ: Prentice-Hall, 1979; Malone, T.W., *Organizing Information Processing Systems,* Working Paper, Palo Alto Research Center: Xerox, 1982.

[10]Emery, J. C., *Organizational Planning and Control,* New York: MacMillan, 1969.

[11]March and Simon, *Organizations,* 1958; Emery, *Organizational Planning and Control,* 1969; and Malone, *Organizing Information Processing Systems,* 1982.

[12]March and Simon, *Organizations,* 1958.

[13]Emery, *Organizational Planning and Control,* 1969.

[14]Simon, *The Sciences of the Artificial,* 1981.

[15]Ibid.

[16]Emery, *Organizational Planning and Control,* 1969.

[17]Ibid.

[18]Cyert and March, *A Behavioral Theory of the Firm,* 1963.

[19]Emery, *Organizational Planning and Control,* 1969.

[20]March and Simon, *Organizations,* 1958.

[21]This is one of the more tragic and striking examples of the dangers of optimizing within narrow boundaries.

[22]Emery, *Organizational Planning and Control,* 1969.

[23]Fallows, J., *National Defense,* New York: Random House, 1981; Coulam, R., *Illusions of Choice: The F-111 and the Problems of Weapons Acquisition Reform,* Princeton: Princeton University Press, 1977.

[24]Emery, *Organizational Planning and Control,* 1969.

[25]Ibid.

[26]Simon, *The Sciences of the Artificial,* 1981, and Emery, *Organizational Planning and Control,* 1969.

[27]Emery, *Organizational Planning and Control,* 1969.

[28]Ibid.

[29]Williamson, O. E., and Ouchi, W. G., "The Markets and Hierarchies Program of Research," in A. H. Vandeven and W. F. Joyce (eds.), *Perspectives on Organizational Design and Behavior,* New York: John Wiley, 1981.

[30]The practice of holding specialty resources aside to be

temporarily attached to mission subunits is an effective strategy for overcoming this problem. The main cost of this strategy is that it increases the degree of involvement by the higher level commander. See Emery, 1969 and Simon, 1981.

[31]It is not clear, however, that functionally structured organizations should be any less prone to subunits parochialism. Simon, *Administrative Behavior,* 1976.

[32]Coulam, Crecine, Fischer, Salomone, *Problems of Command and Control in a Major European War,* Marina del Rey: The Analytical Assessments Corporation in conjunction with Carnegie Mellon University, AAc-TR-21220/83, prepared for the Director, Office of the Secretary of Defense/Net Assessment, pp. 67-73, (hereafter referred to as Crecine, et al.); and Coulam and Fischer, "An Organizational Theory Approach to the Assessment of Military Command and Control Capabilities," in R. Coulam and R. Smith, (eds.), *Advances in Information Processing in Organizations,* Volume II, Greenwich, CT: JAI Press, 1985.

[33]Crecine, et al., pp. 47.

[34]Crecine, et al., pp. 72-73.

[35]Ibid., pp. 72.

[36]Crecine, et al., pp. 50.

[37]Ibid., pp. 74.

[38]Ibid., pp. 75.

[39]Ibid., pp. 51.

[40]Ibid., pp. 75-76.

6

Human Decisions in an SDI Command and Control System

Jean MacMillan and Joseph G. Wohl

Introduction

The Strategic Defense Initiative (SDI) program is a technologically complex and ambitious undertaking. Identifying the appropriate role for human decision making in the resulting system architecture is a challenging task. Although there has been a consensus from the beginning of the program that human involvement in strategic defense decisions is essential, there are significant practical limits on the role that can be played by humans. The issue is how to define and support the role of human decision making, given the characteristics and limitations of human information processing capabilities. If meaningful human roles are to be established in a Strategic Defense System (SDS), early incorporation of human factors analysis in system design is indispensable.

The purpose of this study was to identify ways in which humans can best make a contribution to the effectiveness of an SDS, based on analysis of potential human roles and activities in system operations. Our goal was to identify requirements and objectives for human performance, and to trace the flow of information supporting human functions within the system. This study focuses on human roles in the SDS Command Center during the peacetime and transition phases of system operations. The command center is the place where the most critical human activities will take place in an SDS. The peacetime and transition phase was selected for study because, even though peacetime activities will form the vast majority of all SDS activities, very little analysis has been devoted to this phase.

The key roles for humans in the command center involve authority, responsibility, accountability, and coordination. Human beings must exercise authority over the system. They must take responsibility for system status and readiness, and for the validity of system inputs. They must also be accountable to higher authority for all actions taken. Finally, they must coordinate with each other in order to carry out the system objectives.

Our task was to analyze peacetime and transition functions for selected SDS subsystems. We have chosen to focus our analysis on situation assessment functions in the command center. Situation assessment involves the weighting and fusing of many different types of information, including operational information about the functioning of the system and intelligence information about the state of the world, into an overall picture. Situation assessment draws on the unique capabilities of human beings to merge and interpret information, and provides a fertile area for investigating effective human roles in SDS.

The analysis of control functions in complex systems, and the allocation of system functions to humans and machines, have been topics of considerable interest in recent years. A methodology for functional allocation has been proposed for nuclear power plants (Pulliam et al., 1983) and generalized for application to aerospace systems (Pulliam and Price, 1985). The first stage in this methodology is the identification of those functions for which automation is mandatory, and those functions that must be performed by humans. A series of considerations and suggestions is then given for allocating those functions that fall into the grey area between mandatory automation and mandatory human responsibility.

Our analysis concentrates on the first step of this allocation process by identifying the key decisions that must be made by human beings.

We then consider the support requirements for these decisions. What must the decision maker know in order to make the decision, and what are the sources of this information? The analysis drew on published materials about SDI command and control functions (SDI BM/ Working Group for Standards, July 1987), as well as SDI architecture reports prepared by contractors. We also analyzed existing system architectures for systems with some similarity to SDS, especially NORAD architectures.

Situation Assessment Functions

Situation assessment is the process of integrating information from multiple sources to form a coherent picture. The SDS command center must assess both its own internal situation and the external world situation, and combine them into a total situation assessment that supports critical decisions.

The data for internal situation assessment come from monitoring the Strategic Defense System. This monitoring produces detailed data on system functioning over time, and is the first step in identifying any problems in sensor capability. Based on this health and status data, the command center must assess current and projected sensor capabilities, identify adjustments and fixes needed, estimate the time needed to make these changes, and make a set of critical decisions.

The data for external situation assessment come from a variety of intelligence sources as well as from higher authority (e.g., DEFCON level). External situation assessment is the process of fusing data from multiple sources into an overall picture of the current world situation. As part of this process, indicators and warnings are generated that reflect world tension levels and enemy activities. Another activity of external situation assessment is the estimation of possible time lines for enemy actions. Based on intelligence data, what is the earliest time that an attack might occur?

Assessments of the internal and external situation must be combined with information about system readiness requirements to make an overall assessment of the situation. Any problems with sensor capability must be evaluated in the context of world events. Even minor problems with the surveillance system may be of great concern if DEFCON level is high, and

certain critical indicators and warnings are present. The time needed to adjust or fix sensors must be evaluated in comparison to the estimated time of earliest enemy actions. For example, sensor problems that can be adjusted or repaired before the earliest estimated time of an enemy attack are not as serious as problems that cannot be fixed within the estimated time available. Based on the total internal and external situation, the command center must evaluate the adjustments that can be achieved within the estimated time available and determine the resulting projected defense capabilities.

Five critical decisions that must be made by humans in the command center during peacetime and transition were identified, based on the total situation assessment:

Adjust sensor sensitivity thresholds. The sensitivity thresholds of the sensors may be changed so that they react to lower or higher levels of infrared radiation. Higher sensitivity levels are associated with a higher probability of detecting targets, but at the cost of a higher probability of false alarms. Lower sensitivity levels are associated with a lower probability of false alarms, but a lower probability of detection.

Initiate surveillance sensor cool down. Sensors designed to detect and track targets during their midcourse phase must be able to sense the infrared radiation of a target against a cold background. These sensors must be cooled down to temperatures below the temperature of their satellite environment through stored cryogen, refrigeration, or some other cooling method. This cooling process takes time, so sensor cool down must be initiated before sensor function is needed. However, cooling capacity is limited, so cooling down the sensors shortens their useful lifetime. This may represent an important "false alarm" penalty.

Enable weapons (as authorized by higher authority). Fire control systems must be enabled before they are needed, and this process will not be instantaneous. (Note that the sensors guiding the weapons must be cooled down in the same way as the surveillance sensors.)

Recommend a change in DEFCON level. Although the SDS Command Center will not have the authority to change the DEFCON level, it may make recommendations to higher authority about such changes and provide supporting information.

Recommend or select Rules of Engagement (ROEs) options. Rules of engagement specify actions to be taken in response to different situations. Based on their situation assessment, the SDS command center may decide to recommend a change to the ROE option currently in effect. ROE changes may alternatively be selected within the command center, depending on operating doctrine yet to be determined.

Of all the activities to be carried out in the command center, these five require drawing conclusions about the meaning of information and making decisions based on those conclusions. Total situation assessment requires the complex integration and interpretation of data from multiple sources; much of the data will not be in precisely defined quantitative form. This is an area in which human performance is still consistently superior to machine performance. Finally and most importantly, the decisions based on that assessment *must* be made by human decision makers because humans must take final responsibility for the system's actions. Thus, there is an essential human role in those command center functions.

Flow of Information Among Functions

Figure 1 shows the flow of information among the functions that support situation assessment. Some major themes emerge from the analysis of information flow. Data are converted into information through successive analysis and aggregation. Data on systems operation are processed at successive levels of detail, starting with the detailed raw data about the individual operations of every sensor in the system, and leading up to a global overview of system status based on aggregated data.

The functions progress from maintaining the system, at the lowest level of aggregation, to monitoring system operations, at a higher level of aggregation, to assessing system status, at the highest level of aggregation. The assumption is made that data are analyzed, aggregated, and passed upward at each successive level. For example, a person at the lowest level might see extensive data on an individual sensor, while a person at the highest level would see aggregated data for all sensors in the system.

Evidence of enemy ASAT action might be detected at any point in this aggregation process. An enemy action might be noticed at the lowest level for an individual sensor, or it

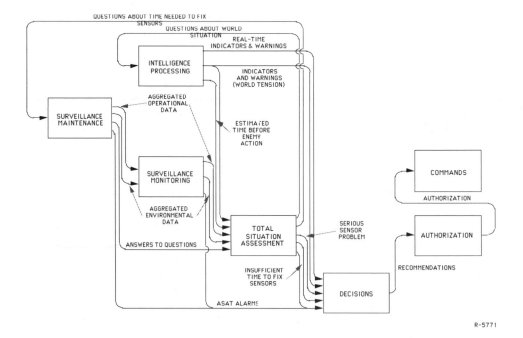

Figure 1. Overview of information flow in situation assessment.

might not be detected until the data were aggregated and a global pattern could be seen. At the point that an ASAT action is noticed, we have assumed that it would be reported directly to the decision function at the command center.

External situation assessment takes place under the intelligence processing component, that produces real-time indicators and warnings that are passed directly to the decision component, and more traditional indicators and warnings that are passed to both the situation assessment component and the decision component.

Information about both the internal and external situation comes together in the total situation assessment component. Questions about the world situation are referred back to intelligence processing, and questions about the time required to fix any sensor problems are referred back to surveillance maintenance. The total situation assessment component of the model passes on any sensor problem judged to be serious (when evaluated in the context of the world situation) to the decision component. If there is insufficient time to fix the sensors before the earliest estimated enemy action, this also generates a message to the decision component.

Another theme of information flow is the presence of an authorization loop for critical decisions. Once a decision has been made, it is passed to higher authority as a recommendation. If authorization is received, then a command is issued. Authority structure has not been defined officially for the SDS command center. However, we assume that at least some of the decisions made in the command center would require prior authorization by higher authority.

Model of Situation Assessment

Events in the real world are linked to decisions in the command center through the flow of information. The architecture of the command and control system and the organization of the command center will determine the path of this information flow. Although the system architecture and the organization for the SDS command center have not yet been precisely defined, we were able to trace the flow of information among command system functions by making a set of reasonable assumptions about system architecture and organization.

This paper analyzes the effects of varying some of these assumptions.

We developed a situation assessment model to support sensitivity analysis of the time required to make decisions under different assumptions about the architecture and organization of the command center. The point of the model is not to provide absolute answers, but to allow meaningful comparisons. It provides a framework for asking questions about the effect of changes in such command center design variables as centralization of functions, lines of communication, distribution of expertise, and authority structure.

The time required to make a decision following an event was selected as the central variable for the sensitivity analysis because of the critical time requirements for SDS decisions. Minutes and seconds, rather than hours, will be the meaningful units for analyzing SDS command center reaction times. Obviously, the accuracy of decisions will also be extremely important, and a tradeoff can be expected between decision accuracy and reaction time. This issue is not dealt with specifically in our analysis because we lacked the detailed information needed for a meaningful analysis of decision accuracy, but it is an essential area for further work as the SDS command and control system is more precisely defined.

The information flow structure shown in Figure 1 was converted into a Stochastic, Timed, Attributed Petri Net (STAPN) model of situation assessment. STAPNs are an extension of the network models invented in the early 1960s by Petri to characterize concurrent operations in computer systems (Peterson, 1961). STAPN models are well-suited for modeling information flows and decision processes because they can be used to keep track of the information that must be present before a particular decision can be executed, the stochastic time delays that are associated with decisions, and the probabilistic outcomes of decision processes (Wohl and Tenney, 1986).

The STAPN model was implemented on a Macintosh computer using ALPHATECH's Micro-Modeler software. This software supports the construction of STAPN models, the assignment of time parameters to transitions in the model, and the assignment of priorities or probabilities to transitions. After the model is constructed, it may be executed, and performance measures such as response time and delays may be computed.

Sensitivity Analysis

The purpose of the model is to analyze the elapsed time between events and decisions under different assumptions about command center architecture and organization. In order to perform this sensitivity analysis, we must assign numerical time values to each of the transitions in the model that represents a process or activity that will consume time. Ideally, these time parameters should be based on empirical data about the time actually taken to complete those tasks. Because the design of the SDS command center is still in a preliminary stage, such data are not available. Instead, we have made a set of arbitrary assumptions about possible times that seemed reasonable, and then varied them to test the sensitivity of the overall response time to the processing times for individual components.

Response time, in this context, refers to the time that elapses between the time that raw data are received by the system, and the time that a command is issued based on those data. For example, in the case of an ASAT action, how much time elapses between the receipt of data from which this attack could be detected and the issuing of a command to cool down the sensors based on the detection of the attack? Of course, some time will elapse between the time that an event occurs and the time that the raw data reflecting that event are received by the system, but this is outside the scope of our model.

The first step in the sensitivity analysis is to establish a set of reaction times that are of interest. We want to examine the links between commands based on critical decisions and the events that generate them. Our approach was to develop a set of "scenarios" of possible events and identify the critical decisions that might be triggered.

The four scenarios and the critical decisions triggered by each are shown in Table 1. The first scenario is simply the detection of a real-time indicator and warning. This is assumed to trigger four critical decisions: cooling down the sensors, changing sensor sensitivity thresholds, recommending a DEFCON change, and enabling the weapon system.

The second scenario involves sensor capability problems: indicators and warnings show a high state of world tension, the DEFCON level is high, and a serious sensor capability problem has emerged. The decisions triggered by this scenario are cooling down of the sensors, changing thresholds, and recommending a shift in ROE option.

Scenario 3 adds a time "window" constraint to scenario 2: There is not enough time to fix the sensor problem before the estimated time of the earliest enemy attack. This is assumed to trigger the additional decisions to recommend a DEFCON change, and to enable the weapon system.

	CRITICAL DECISIONS				
SCENARIOS	Cool down sensors	Change sensor threshold	Recommend DEFCON change	Recommend ROE shift	Enable weapons
1. A real-time I&W occurs.	X	X	X		X
2. I&W show high world tension, DEFCON is high, and a serious sensor capability problem exists.	X	X		X	
3. I&W show high world tension, DEFCON is high, a serious sensor capability problem exists, and there is not enough time to fix it before the estimated time of earliest enemy action.			X		X
4. An ASAT action occurs, and is detected:					
• early, during maintenance;	X	X	X		X
• later, during monitoring;	X	X	X		X
• not until assessment, when operational, environmental, and intelligence data are combined.	X	X	X		X

Table 1. Links between scenarios and critical decisions.

The fourth scenario deals with the detection of ASAT actions. Three possible detection points are included: early detection, as raw data for individual sensors are processed during surveillance maintenance; later, as more aggregated data are processed during surveillance monitoring; and still later, during situation assessment, when global operational and environmental data are analyzed along with intelligence data. The decisions triggered by the detection of an ASAT attack are assumed to be the same as those triggered by real-time indicators and warnings: cooling down the sensors, changing the sensor thresholds, recommending a DEFCON change, and enabling the weapon system.

The model may now be used to estimate the total aggregate time that would elapse between initial data receipt and the issuing of different types of commands for the scenarios shown in Table 1. (Note: As computed by the model, the total aggregate time between a given triggering event and the issuing of commands depends directly on the assumed time delay for each individual activity leading to those commands. Thus, the arbitrariness of the individual time delay parameters is reflected in all response time predictions.)

The model with its present assumptions provides a tool for analyzing how some of the factors in command and control system architecture and the command center organization might affect reaction times. Variables that could be manipulated in organizational and system design include the centralization of functions (geographically or organizationally), the coordination between the command center decision functions and the intelligence processing and surveillance maintenance functions, and the authority structure for approving command decisions. The remainder of this section explores the effects of shifts in these factors on reaction times.

The Centralization of Functions. Transmittal and communications times are pervasive throughout the model. This includes the time needed to pass information from the intelligence processing function to the decision function; to transmit aggregated data from the surveillance functions to the situation assessment function; to transmit questions back to surveillance and intelligence processing; to transmit requests for authorization to higher authority; and for authorization to be transmitted back to the command function.

A number of factors could increase or decrease transmittal and communication times. The design of the communications system is critical, as is the organizational design of the command center. If functions are performed by the same individual, or by individuals within the same part of the organization or located in the same area, then communication time may be much less than if the same functions are performed by a group of dispersed individuals. For purposes of the sensitivity analysis, two transmittal times were tested: 0.25 minutes and 2 minutes. Table 2 shows the effect on

CRITICAL DECISIONS

SCENARIOS	Cool down sensors	Change sensor threshold	Recommend DEFCON change	Recommend ROE shift	Enable weapons
1. A real-time I&W occurs	4.0 **7.5**	4.0 **7.5**	7.5 **11.0**		4.0 **7.5**
2. I&W show high world tension, DEFCON is high, a serious sensor capability problem exists.	41.5 **48.5**	39.0 **51.0**		43.0 **50.0**	
3. I&W show high world tension, DEFCON is high, a serious sensor capability problem exists, and there is not enough time to fix it before the estimated time of earliest enemy action.			68.5 **79.0**		69.0 **79.5**
4. An ASAT action occurs, and is detected:					
• early, during maintenance;	13.0 **16.5**	13.0 **16.5**	16.5 **20.0**		13.0 **16.5**
• later, during monitoring;	24.25 **29.5**	24.25 **29.5**	27.75 **33.0**		24.25 **29.5**
• not until assessment, when operational, environmental, and intelligence data are combined.	35.5 **42.5**	35.5 **42.5**	39.0 **46.0**		35.5 **42.5**
Note: Numbers in regular type are based on a transmittal time of 0.25 minutes. Numbers in **bold** type are based on a transmittal time of 2 minutes.					

Table 2. Elapsed time in minutes before command is issued under two different assumptions about transmittal and communication times.

CRITICAL DECISIONS

SCENARIOS	Cool down sensors	Change sensor threshold	Recommend DEFCON change	Recommend ROE shift	Enable weapons
2. I&W show high world tension, DEFCON is high, a serious sensor capability problem exists.	41.5 **57.0**	39.0 **54.5**		43.0 **58.5**	
3. I&W show high world tension, DEFCON is high, a serious sensor capability problem exists, and there is not enough time to fix it before the estimated time of earliest enemy action.			68.5 **84.0**		69.0 **84.5**

Note:
Numbers in regular type are based on the assumption that intelligence processing takes place in parallel with processing of operational data.
Numbers in **bold** type are based on the assumption that intelligence processing does not begin until questions are generated from the processing of operational data.

Table 3.   Elapsed time until command is issued under two different assumptions about coordination with intelligence processing.

reaction times if the time for transmittal and communication is increased from 0.25 minutes to 2 minutes throughout the model.

Obviously, an increase in transmittal time will increase reaction times. The more functions involved in a scenario, the larger this effect will be. The increase in reaction time for scenario 1, based on real-time indicators and warnings, is 3.5 minutes (calculated as the difference between the reaction times under the two transmittal time assumptions). The increase for scenario 2, based on sensor capability asssessment, is 7 minutes. For scenario 3, which involves analysis of the time available before enemy actions, the increase is 10.5 minutes. For scenario 4, the increase in reaction time for ASAT actions detected early is 3.5 minutes. As more processing is required, the increased communication time will have more effect. If an ASAT action is not detected until situation assessment, the increase in reaction time due to slower communication is 7 minutes.

Coordination with Intelligence Processing. The reaction times shown in Table 2 assume that the processing of intelligence data and the generation of indicators and warnings takes place in parallel with the processing of data about surveillance system operations and local satellite environments. If intelligence data and operational data are received at the same time, then indicators and warnings will have been

prepared and passed to the total situation assessment function by the time that aggregated operational data are received. In other words, the time spent generating indicators and warnings of world tension under intelligence processing is not part of the critical path that drives reaction times for scenarios 2 and 3 in Table 2.

This assumption of parallel processing may not be realistic. For example, a structure in which intelligence processing is driven by questions generated during situation assessment might be more similar to current practices. Under this assumption, the generation of indicators and warnings from intelligence data would not start until sensor capability problems had been identified and questions had arisen about the seriousness of those problems in the context of world tension. If the model is altered to reflect this structure, then the time needed to generate indicators and warnings will become part of the critical path that determines reaction times and will substantially increase them.

Table 3 shows reaction times (in bold type) for scenarios 2 and 3 (which depend on the generation of indicators and warnings) under the revised assumption about intelligence processing. If the generation of indicators and warnings is reactive, rather than proactive, the reaction time under both scenarios 2 and 3 increases by 15.5 minutes. That is, waiting until sensor problems arise to begin analyzing intelligence data, substantially increases reaction time. Continual, on-going analysis and evaluation of intelligence data, that is, proactive fusion, supports much faster reactions.

Coordination Between Surveillance and Command Functions. The reaction times shown in Tables 2 and 3 assume that questions about the time needed to make sensor adjustments or fix sensor problems are generated during total situation assessment, and referred back to sensor maintenance experts in the surveillance maintenance function. These experts then transmit answers back to situation assessment. This process is part of the critical path for decisions in scenario 3.

The assumption that a round of questions and answers must occur between the surveillance maintenance and the situation assessment functions may be incorrect for several reasons. First, the individuals responsible for situation assessment may know how long sensor repairs and adjustments will take, without having to consult the sensor maintenance staff. Also, the sensor maintenance staff may anticipate the needs of situation assessment, and may pass information about the time needed to make adjustments at the same time they transmit information about problems in sensor capability.

In either case, the result will be faster reaction times for those actions triggered by the assessment that not enough time is available, as shown in Table 4. Reaction time is decreased by 5.25 minutes under the assumption that estimates of the time needed for sensor repairs can be made by the individuals performing situation assessment, without referring back to surveillance maintenance.

Authority Structure. The structure of the model (See Figure 1) provides an authorization loop for all decisions made in the SDS command center. For each command, a message must be transmitted to higher authority, an authorization decision must be made, and the authorization must be transmitted back to the command center before the command can be issued.

Obviously this process consumes time, and it is in the critical path for all decisions. However, these assumptions are dependent on the not-yet-determined authority structure of the SDS command center, and reasonable alternatives may be generated. For example, perhaps decisions that involve the surveillance system, such as sensor cool down and threshold changes, need not be approved by higher authority.

Table 5 shows the decrease in reaction time from removing the authorization loop for sensor decisions. Authorization time is on the critical path for sensor decisions, so the removal of the time needed for transmission and authorization has a direct effect on reaction times.

CRITICAL DECISIONS

SCENARIOS	Cool down sensors	Change sensor threshold	Recommend DEFCON change	Recommend ROE shift	Enable weapons
3. I&W show high world tension, DEFCON is high, a serious sensor capability problem exists, and there is not enough time to fix it before the estimated time of earliest enemy action.			68.5 **63.25**		69.0 **63.75**

Note:
Numbers in regular type are based on the assumption that the total situation assessment function must refer questions about the time needed to make sensor adjustment back to the surveillance maintenance function.
Numbers in **bold** type are based on the assumption that the total situation assessment function does not refer to questions about the time needed to make sensor adjustments back to the surveillance maintenance function.

Table 4. Elapsed time before command is issued under two different assumptions about coordination between surveillance and command functions.

CRITICAL DECISIONS

SCENARIOS	Cool down sensors	Change sensor threshold	
1. A real-time I&W occurs	4.0 **2.5**	4.0 **2.5**	
2. I&W show high world tension, DEFCON is high, a serious sensor capability problem exists.	41.5 **36.0**	39.0 **33.5**	
4. An ASAT action occurs, and is detected:			
• early, during maintenance;	13.0 **11.5**	13.0 **11.5**	
• later, during monitoring;	24.25 **22.75**	24.25 **22.75**	
• not until assessment, when operational, environmental, and intelligence data are combined.	35.5 **34.0**	35.5 **34.0**	

Note:
Numbers in regular type are based on the assumption that sensor commands must be approved by higher authority.
Numbers in **bold** type are based on the assumption that sensor commands do not require approval by higher authority.

Table 5.   Elapsed time before sensor commands are issued, with and without the need for approval by higher authority.

Removing the need for authorization cuts 1.5 minutes for decisions based on real-time indicators and warnings and ASAT alarms, and 5.5 minutes for decisions based on sensor capabilities.

Conclusions

The sensitivity analysis based on the model identifies issues that must be addressed in evaluating designs for an SDS command center. The model provides a structure for analyzing the links between functions and the implications of those links for the overall reaction time. Any design for an SDS command center must take into account the relationship between information flow and reaction speed.

The centralization of functions and the speed of communications within the command center will have a pervasive effect on the speed with which decisions can be made and commands issued. Situation assessment in the command center will require the fusion of information from multiple sources, and the speed of the slowest link will determine the speed of the overall assessment.

The extent to which questions and answers pass back and forth between functions is another major factor in reaction time. Interchanges take time, and if situation assessment can be done without the need for referring questions back to intelligence processing or surveillance maintenance, this will speed reaction time con-

siderably. The elimination of such interchanges will require either that the staff involved in situation assessment be extremely knowledgeable about the intelligence situation and the functioning of the surveillance system, or that staff in those areas anticipate the needs of the situation assessment staff and pass on exactly the information that is needed.

The integration and coordination of intelligence analysis with situation assessment is a key factor in reaction times. If intelligence analysis is done only on request, after serious sensor capability problems have been identified, this will slow reaction times by a substantial amount. Continual, on-going intelligence analysis that anticipates the needs of situation assessment in the SDS command center will lead to faster reaction times.

Authority structure is another important element of reaction time. Authorization loops are part of the critical path for all of the decisions in the model. A reduction in the time needed to transmit intended commands to higher authority and to receive authorization to proceed will reduce command center reaction times by an equal amount.

This sensitivity analysis is intended only as an illustration of the kinds of questions that must be evaluated in assessing command center designs. The basic idea of the STAPN model is to establish linkages among functions and to identify critical paths of information flow leading up to decisions. As plans for an

SDS command and control system are further developed, it should become possible to assign more realistic values to time delays in the system and reach more exact conclusions about reaction times under different designs.

References

BBN Laboratories Inc., An Air Force Advanced Engineering Program for SDI BM/ Human Factors: Final Report, October, 1986.

Peterson, J. L., *Petri Net Theory and the Modelling of Systems*, Prentice-Hall: Englewood Cliffs, New Jersey, 1981.

Pulliam, R. and H. E. Price, *Automation and the Allocation of Functions Between Human and Automatic Control: General Method*, Air Force Aerospace Medical Research Laboratory, February, 1985.

Pulliam, R., H. E. Price, J. Bongara, C. R. Sawyer and R. A. Kisner, *A Methodology for Allocating Nuclear Power Plant Control Functions to Human or Automatic Control*, Union Carbide Corporation, August, 1983.

SDI BM/ Working Group for Standards, *Functional Decomposition*, July, 1987.

Wohl, J. G. and R. Tenney, "Integrated Analysis Techniques for Command, Control, and Communications Systems, Volume I: Methodology," Technical Report TR-293, Alphatech, Inc., November, 1986.

7

Normative Models for Capturing Tactical Intelligence Knowledge

Rex V. Brown

Introduction

The Decision Aiding Task

The contribution of tactical intelligence to the success of a future military engagement depends critically on staff knowledge we cannot take for granted. It may be factual (e.g., about enemy practices) or procedural (e.g., how to manage intelligence collection and analyze a situation). The task is to extract the wealth of fragmented knowledge available on the art of tactical intelligence and synthesize it in a form that can be used effectively in the field—for instance, by an army division G-2.

This knowledge is resident in a variety of overlapping humans and documents, including military professionals, substantive and methodological experts, as well as doctrinal/training literature and research studies. However, it is incomplete, often conflicting, liable to rapid obsolescence (especially after hostilities begin) and there is no way of assuring that it will be at the fingertips of intelligence personnel when they need it in preparation for a battle—much less in mid-combat.

Special Requirements

Computerized "knowledge-based expert systems" have emerged in recent years as a promising artificial intelligence approach to solving this type of problem. However, the type of knowledge we are concerned with here has a number of distinctive features that may radically affect the appropriate mode of elicitation or the format of representation.

It involves combat elements (situations, threats, intelligence data, and collection options), which cannot be sharply envisaged far in advance (i.e., when an expert system would be developed). By contrast, the success stories of AI expert systems have tended to be for relatively well-structured problems, such as medical diagnosis, oil exploration, and engineering design. In addition, since intelligence knowledge refers to a moving target, military practice is perpetually changing (by contrast to medical diagnosis of human ailments).

Furthermore, this knowledge is institutional and dispersed: we are not interested in any individual's knowledge per se. Indeed there may not be any single expert whose knowledge, even about a specific task, encompasses enough of what is available that we should rely on his expertise alone. No one has yet had direct practice at the tasks we are preparing for (e.g., in World War III).

All of this argues for a framework for eliciting and representing knowledge which can accommodate a variety of human and other sources, can be readily updated, and, most important, can assimilate last minute input from field personnel who have no special training. This suggests categorizing knowledge at three levels of generality, typified by: a field manual, referring, for instance, to any U.S.-Soviet conflict in Central Europe; the intelligence preparation of a specific European battlefield; and judgments developed in the course of a particular engagement.

The PDA Approach

We propose to adapt a well-established normative approach, where knowledge is represented in a logically coherent structure; specifically as a personalized decision analysis (PDA) model, consisting of a coherent set of quantitative measures of probability and utility (Watson and Buede, 1988). Two familiar variants of a PDA model are choice evaluation through subjective expert utility and inference through Bayesian updating (Barclay, et al., 1977).

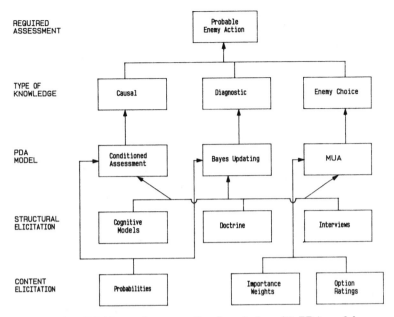

Figure 1. Eliciting and representing knowledge with PDA models.

The knowledge that can be represented in this way is of two types. Structure (for example, what enemy intent gives rise to what intelligence indicators) is reflected in the form and labeling of the model. Content (such as the extent to which an indicator supports a hypothesis about intent) is reflected in probability and other numbers.

Generally, the PDA approach to knowledge elicitation and representation (KER) is practically and conceptually well-advanced, and indeed mature software is available for several variants.

However, much development and testing is needed before an operational KER technology is made available. In particular, it is not trivial to design a PDA model so that it reasonably fits both the form in which the knowledge is originally found (e.g., the experts' cognitive structures) and the use for which it is intended (e.g., output usable by a typical G-2).

Progress to Date

Currently, DSC is engaged in a major project for the Army Research Institute to develop knowledge elicitation methodology and the underlying psychological theory, primarily for Army tactical intelligence purposes (Cohen and Leddo, 1987, in press; Leddo et al, 1988; Cohen and Leddo, in preparation.).

One of many strands of that project has been to explore normative PDA models as a vehicle for certain types of knowledge. The KER techniques we address here are the result. We have adopted a "build-test-build-test" approach to developing those techniques and have had several rounds of trying them out experimentally in Army contexts, using active service Army intelligence specialists as subjects (Brown, 1989). There are now clear tests of feasibility and practical promise which we believe are ready for building into software prototypes and for field testing.

The focus has been on two common KER tasks discussed below. The first involves a battlefield aid to support selected common intelligence analysis tasks, such as predicting the location of an enemy attack. The second task is to foresee the possible outcomes of a combat situation, through a field-based training aid.

Predicting Enemy Action

When probable enemy course of action or intent is to be assessed, three kinds of knowledge and reasoning can be brought to bear: causal, diagnostic, and projecting the enemy decision process. Each of these can be represented by a PDA model whose structure and content can be developed independently (See Figure 1).

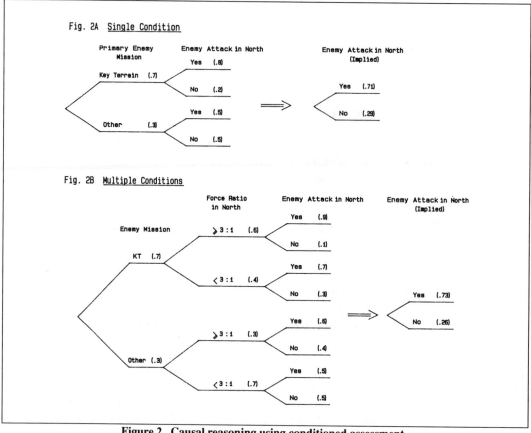

Figure 2. Causal reasoning using conditioned assessment.

Conditioned assessment models (sometimes represented as influence diagrams) capture causal judgments of the "it all depends" type (e.g., location of enemy attack depends on his mission). Within this framework, predictions of enemy intent (e.g., mission) will often be anchored to second-guessing the enemy's decision process, for example, in the form of a multiattribute utility model. Updating probabilistic assessments in the light of diagnostic intelligence developments can be captured in Bayesian updating analysis.

Figures 2, 3, and 4 illustrate each of the above techniques with an example addressing the same KER task. That is, as part of an Army division's Intelligence Preparation of the Battlefield, how probable is it that the enemy will attack in the North? The setting is, for example, a delineated sector of the Fulda Gap region in western Germany after the first few days of a Soviet offensive. Detailed applications of these techniques in this hypothetical scenario, using active Army intelligence personnel as elicitation subjects, are illustrated in Brown (1989).

Conditioned Assessment

The conditioned assessment model shown in Figure 2 has a simple but powerful logic for exploiting the case where the expert's (or other source's) knowledge bears most naturally on one or more causes of the event of interest. When the relevant conditional and unconditional assessments are assigned (using well-established probability elicitation techniques), the required probabilities are straightforwardly calculated. Multiple models can be constructed for the same required assessment, using different conditioning variables, singly or (as in Figure 2b) in combination. Influence diagrams can be used to help organize the linkages when they become complex.

The judgment and data embodied in such PDA models can be logically decomposed into separable sets of inputs supplied by the same or different sources. In particular, model structure and default content numbers can be developed ahead of battle, using substantial effort and skill, which may not be available in the heat of battle. For example, in the conditioned assess-

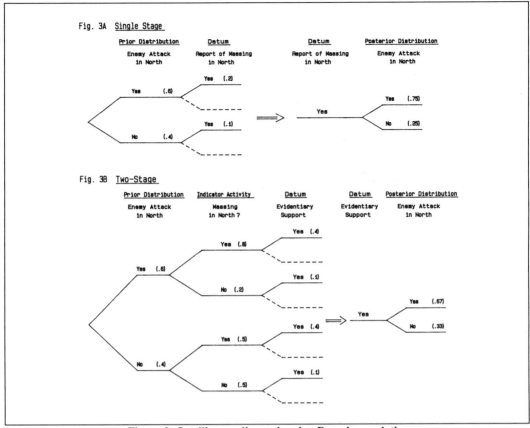

Figure 3. Intelligence diagnosis using Bayesian updating.

ment scenario, a menu of typically appropriate model structures would be developed ahead of time (such as mission and/or force ratios as determinants of location of attack); as well as any judgment or data on the corresponding conditional probabilities (e.g., on how likely the enemy is to attack when he has a force-ratio advantage of less than 3:1).

Much of the practical value of storing knowledge in this form will depend on being able to complete the model during combat, when the assessment is needed. This requires a user-friendly input/output interface and an adequately trained staff. This is a powerful reason to keep the structure simple.

Bayesian Updating

Figure 3a shows the simplest form of a Bayesian updating model, based on the well-known prior-posterior paradigm, where the required assessment is periodically updated as new data is received. The BAUDI program developed for ARI is designed to be used in real time to handle reports of any degree of complexity, provided only that there is someone to assess its diagnosticity, in appropriate conditional probability terms. In this form, all the elicitation is done during battle, and there is little scope for storing knowledge of enduring interest.

However, a two-stage model of the type shown in Figure 3b permits partitioning out knowledge to be stored ahead of time. Specifically, a required assessment of a given kind (e.g., enemy attack in a given location) can be related (as in the relevant field manuals) to a broadly defined indicator event (e.g., massing of forces) of general applicability. (This kind of linkage is already presented in doctrinal field manuals). Diagnostic generalizations can be stored as conditional probabilities (default likelihood functions), which can be overridden when necessary.

On the battlefield, incoming intelligence only needs to be characterized in terms of how much evidentiary support it provides for the indicator event (typically a variable rather than a simple yes-no as in this simple example).

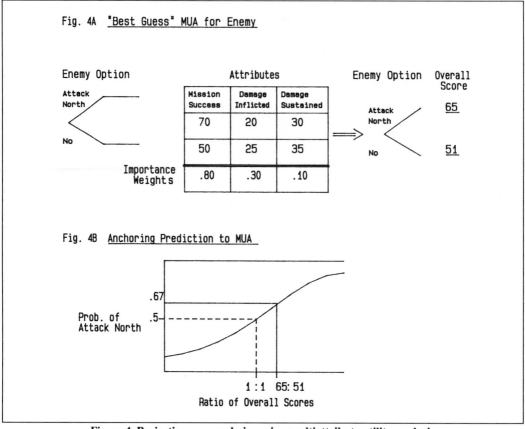

Figure 4. Projecting enemy choice using multiattribute utility analysis.

Conditional probabilities (likelihoods) of evidentiary support given indicator activity could also be given default values in advance. The prior probability (e.g., of enemy attack in this location) would be supplied, on the spot, either by direct judgment, or as output from another exercise, such as conditioned assessment.

Multiattribute Utility Analysis

The third PDA approach, illustrated in Figure 4, is to capture knowledge about how the enemy will make the choice in question. The ways of doing this are as varied as there are PDA choice models, but again there are default models which can be developed to draw on. Multiattribute Utility Analysis (MUA) models are promising, often using standard attributes, such as those shown in Figure 4a.

A simple and commonly used additive model will usually work well. Using an expert's "best guess" at how the enemy will weight each attribute, and a score for each of the options in question, an overall score for each such option can be calculated. (Default

importance weights would be stored in advance.) These comparative scores need to be converted into probabilities that the enemy will actually adopt each option (since we cannot be sure he will act as we project). This could be done by direct judgment on the battlefield or via a prestored function, such as that shown in Figure 4b. A plausible property for such a function is that if the calculated scores are equal, so are the probabilities (but this does not logically have to be the case).

Plural Evaluation

In general, a given assessment problem can be formulated in different ways corresponding to different "experts" or alternative ways of looking at an assessment by the same expert. These can be pooled or reconciled, and progressively added to or refined, as more knowledge becomes available (Brown and Lindley, 1986).

Figure 5 shows, in a highly schematic form, the elements of an integrated KER strategy using the four forms of PDA (including plural

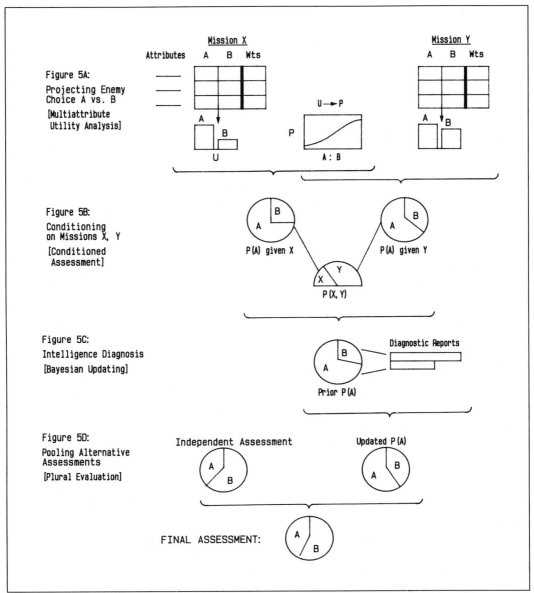

Figure 5. Integrated PDA for predicting enemy action (alternative actions: A vs. B).

evaluation), and indicating some of the main aspects of supplying structure and content in each case. First, the enemy's probable choice under varying hypotheses about his mission is evaluated (Figure 5a). Then these probabilities are combined based on the probabilities of the missions Figure 5b). The resulting probability is updated in the light of incoming intelligence reports which have different diagnosticity for each action (Figure 5c). Finally, the updated probability is pooled with any alternative perspective on the same probability, often including an intelligence officer's direct intuition (Figure 5d).

The four modules can be uncoupled from each other such that their inputs can be supplied either by direct judgment or as output from another module.

Discussion

The intent of PDA models generally is to outdo not emulate experts' knowledge, which is basically what typical AI expert systems set out to do. (An analogy would be developing an airplane versus. copying a bird, if the object is to fly).

PDA models assure that conclusions, whether of decision or choice, are logically coherent and can be reviewed and refined, piece by piece, from a multitude of sources (human and other). Moreover, they can be made to conform, at least in essence, to any sound line of reasoning or prescription (and any logical inconsistency can be detected and corrected). For this reason, it is a promising candidate for computerizing military field manuals or doctrine, with the prospect of making them richer and more finely articulated in the form, for example, of expert systems for field work stations.

Such expert systems usually will require progressive fine-tuning in terms of probability, value and importance numbers, right up to the point of use on the battlefield. Until a substantial amount of user engineering has been done, this will represent a nontrivial burden of time and skill on line and staff in the field. However, this is bound to be the case for any computerized "expert system" that purports to replace or enhance military judgment of any but the most routine kind.

A problem with PDA models for expert systems is that the knowledge of any given human may not fit readily into any particular structure. Cognitive structures differ from one individual to another. (This will be a problem with any formal representation of knowledge which seeks to accommodate multiple sources). Moreover, important knowledge may be "lost in the translation" as qualitative perceptions of uncertainty, value, etc. are forced into the numerical form of probabilities, utilities, etc. This is a good argument for making sure that the user of a system can override it with his direct judgment and generally use it as a "collaborator" rather than a replacement.

Key technical issues that still need to be addressed include:

• verbal elicitation protocols for structure and content;

• computer-aided interaction between model and knowledge user (who may also be a source of last minute knowledge on the battlefield); and

• guidelines for selecting modeling approaches to match situations.

Simulating the Outcome of an Engagement

Step-through Simulation

Predicting the outcome of a military engagement, under varying assumptions, is a critical skill and one difficult to capture explicitly, because, in war, possible developments are usually complex and ill-defined. Step-through simulation is a PDA technique for generating realistic sequences of actions and reactions resulting from a specified scenario setting under any desired assumptions (Brown and Ulvila, 1978).

It does so by having the experts whose knowledge is to be tapped play a special kind of computer-aided war game, which calls for probability distributions instead of single events. One or more experts play the parts of distinuishable outcome determinants, such as own forces, enemy, nature, and combat resolution.

The players essentially conduct computer-aided interactive thought experiments in which they anticipate possible developments in sequence. At each step, probability distributions are assessed and sampled, Monte Carlo style. The resulting sequences are treated as a probability sample of all possible futures, reflecting (implicitly) the judgment and imagination of the participating experts. If the outcome of each sequence is characterized by performance measures, such as mission accomplishment and damage to own and enemy forces, then the whole situation can be evaluated in terms of a mean and range of those measures.

The simulation can be used to: evaluate the probable consequences of alternative friendly courses of action; generate representative scenarios for various conventional purposes; provide prebattle training to participants in the simulation; and contribute to a store of surrogate "case studies" from which generalized knowledge can be synthesized.

Current Status

Although the explicit PDA model approach to capturing predictive expertise (presented earlier) is well-established, the step-through approach is still at an early stage of development and implementation. Its most promising uses and potential value in each are still to be established. Interactive software, though primitive, has been developed for this purpose (Ulvila and Brown, 1978). In parallel, paper and pencil trials using Army intelligence specialists have been conducted, which suggests directions for further development (Brown, 1988).

Discussion

Unlike conventional Monte Carlo simulation, step-through simulation avoids having to specify in advance, even implicitly, all contingencies and their conditional probabilities. Unlike conventional war gaming, it limits vulnerability to cognitive myopia and preconceptions, and exercises the full range of possibilities in a controlled fashion.

Step-through simulation partakes of the same basic logic of PDA, and can be used for many of the same purposes—such as prediction and evaluating courses of action. However, its implementation is likely to be always too cumbersome and demanding to be of much use in time- and resource-scarce battle situations. On the other hand, it may have a significant role to play in prebattle analysis of options and prospects.

This may prove to be an excellent way of making rapid use of the kind of knowledge developed after hostilities begin. Such knowledge, being based on experience with the "real thing," may rapidly make obsolete much of the knowledge available in peacetime (which is based, at best, on exercises and the remembrance of wars past). This is the reason why it is so critical, as we have noted, that PDA models, or any other type of expert system developed in peacetime, should have the capacity to assimilate up-to-date knowledge at very short notice.

Acknowledgement

Work on this paper was supported by the Army Research Institute under contract No. MDA903-86-C-0383.

References

Barclay, S., Brown, R.V., Kelly, C.W., III, Peterson, C.R.,Phillips, L.D., and Selvidge, J. Handbook for decision analysis. Report to DARPA. McLean VA: Decisions and Designs, September 1977.

Brown, R. V. "The Use of Normative Models for Eliciting Tactical Intelligence." Report to the Army Research Institute. Reston, VA: Decision Science Consortium Inc., March 1989.

Brown, R.V. and Lindley, D.V. "Plural Analysis: Multiple Approaches to Quantitative Research." *Theory and Decision* 20, 1986,133-154.

Cohen, M.S. and Leddo, J.M. "Knowledge Elicitation Methodology for Tactical Intelligence." Summary report to Army Research Institute. Reston, VA: Decision Science Consortium Inc., in preparation.

Leddo, J. and Cohen, M.S., "Cognitive structure analysis: A technique for eliciting the content of structure of expert knowledge." Proceedings of 1989 AI Systems in Government Conference, to appear.

Leddo, J.M., and Cohen, M.S. "A cognitive science approach to elicitation of expert knowledge." *Proceedings of the 1987 JDL Command and Control Research Symposium*. McLean, VA: Science Applications International Corporation, 1987.

Leddo, J.M., Mullin, T.M., Cohen, M.S., Bresnick, T.A., Marvin, F.F., and O'Connor, M.F. "Knowledge Elicitation: Phase I Final Report," Vols. I & II (Technical Report 87-15). Reston, VA: Decision Science Consortium, Inc., May 1988.

Ulvila, J.W., and Brown, R.V. "Step-Through Simulation. *Omega: The International Journal of Management Science,* 1978, 6(1), 25- 31.

Watson, J.R. and Buede, D.M. *Decision Synthesis.* Cambridge University Press, 1988.

8

Evaluation of Expert Systems in Decision Making Organizations

Didier M. Perdu and Alexander H. Levis

Introduction

Decision making processes require the analysis of complex situations and the planning, initiation, and control of subsequent responses. These activities are done within some constraints on time and accuracy and so that an acceptable level of effectiveness can be reached. The amount of information handled by decision makers is often very large and, in order to maintain performance above a certain level, organizations use decision support systems to help them accomplish their mission. Among them, expert systems with their deductive capability and their ability to handle symbolic concepts have proved to be very useful. The aim of this paper is to show to what extent the use of an expert system modifies the measures of performance of a decision making organization. To allow the use of the analytical framework developed for the study of these organizations, an expert system model using Predicate Transition Nets is developed that can be used for the evaluation of the response time. Expert systems are then studied to assess their usefulness in aiding the fusion of possibly inconsistent information coming from different sources. This assessment is done through the analysis of an application involving a two-decision maker organization facing this problem. Three strategies used to solve this problem are described, one of them involving the use of an expert system. Measures of performance reached for each of these strategies are finally evaluated and compared.

1.0 An Expert System Model Using Predicate Transition Nets

Knowledge Based Expert Systems show properties of synchronicity and concurrency which makes them suitable for being represented with the Predicate Transition Net formalism (Genrich and Lautenbach, 1981). The rules of a knowledge base have to be checked in a specific order depending on the strategy used to solve the problem and on the current facts deduced so far by the system in the execution of previous rules. A model of an expert system using production rules to represent knowledge is presented. Some previous work (Giordona and Saitta, 1985) has addressed the modeling of production rules of a knowledge base using Predicate Transition Nets. The model presented here is different because it incorporates explicitly the control exercised by the inference engine. Fuzzy logic (Zadeh, 1965 and 1983; Whalen and Schott, 1983) is used to deal with uncertainty, and Predicate Transition Nets are used to represent the basic fuzzy logical operators AND, OR and NOT that appear in these kinds of rules. The combination of these operators permits us to represent the dynamical behavior of an expert system.

1.1 Structure of the Expert System

Knowledge Based Expert Systems, commonly called expert systems, are—in theory—able to reason using an approach similar to the one followed by an expert when he solves a problem within his field of expertise. A net model for the most common kind of expert systems, the *consultant expert system*, as described by Johnson and Keravnov (1985), is proposed. Most systems engage in a dialogue with the user, the computer acting as a "consultant," by suggesting options on the basis of its knowledge and the symbolic data supplied by the user. There are three distinct components in an expert system, the knowledge base, the fact base, and the inference engine.

The knowledge base contains the set of information specific to the field of expertise. The knowledge base is a collection of general facts, empirical rules, and causal models of the problem domain. A number of formalisms exist to represent knowledge. The most widely used is the production system model in which the knowledge is encoded in the form of antecedent-consequent pairs or IF-THEN rules. A production rule is divided in two parts :

• A set of conditions (called left-hand side of the rule) combined logically together with an AND or an OR operator,

• A set of consequences or actions (called also right-hand side of the rule), the value of which is computed according to the conditions of the rule. These consequences can be the conditions for other rules. The logical combination of the conditions on the left-hand side of the rule has to be true in order to validate the consequences and the actions.

The fact base, also known as context or working memory, contains the data for the specific problem to be solved. It is a workspace for the problem constructed by the inference mechanism from the information provided by the user and the knowledge base. The working memory contains a trace of every line of reasoning previously used by memorizing all the intermediate results. Therefore, this can be used to explain the origin of the information deduced or to describe the behavior of the system.

The inference engine is used to monitor the execution of the program by using the knowledge base to modify the context. It uses the knowledge and the heuristics contained in the knowledge base to solve the problem specified by the data contained in the fact base. The inference engine selects, validates, and triggers some of these rules to reach the solution of the problem.

Left-hand sides of rules are combined logically together to deduce the truth or falsity of the right-hand side. However, problems to be solved in artificial intelligence rarely involve the distinction between two opposed elements. Two-valued logic (true-false) does not give an accurate representation of the field of expertise because it is not designed to deal with elements whose truth or falsity is a matter of degree. Fuzzy logic allows this and furthermore underlies inexact or approximate reasoning. Instead of assigning a value true or false to each fact, a

number between 0 and 1 is associated. This number, called degree of truth, is equal to the grade of membership in the fuzzy set of the elements verifying a certain propriety. For example, let X be a set of planes, X = {F16, F18, AWACS} and we know that:

F16 flies at mach 1.5,
F18 flies quite fast,
and AWACS has a low speed.

If A denotes the fuzzy set of fast planes in X, A will be:

$$A = \{ (F16, 0.7), (F18, 0.8), (AWACS, 0.1) \}.$$

The numbers (0.7, 0.8, 0.1) are the grades of membership of the different elements of X to the fuzzy set of fast planes and show to what extent these planes belong to the fuzzy set A.

Fuzzy logic has been implemented in the model to combine logically the conditions of the left-hand side of the production rules. The value of a rule or a fact is either unknown or a number, p_i, between 0 and 1, representing the degree of truth (or grade of membership to certain fuzzy set) associated with it. The operators AND, OR, and NOT correspond to operations on fuzzy sets and operate on these degrees of truth as follows:

p_1 AND p_2 $= \min(p_1, p_2)$
Intersection of two fuzzy sets

p_1 OR p_2 $= \max(p_1, p_2)$
Union of two fuzzy sets

NOT p_1 $= 1 - p_1$.
Complement of a fuzzy set.

To simulate the behavior of an expert system, the process of selection and firing of rules done by the inference engine has been modeled when a backward chaining strategy is used. For this strategy, the system tries to support a hypothesis by checking known facts in the context. If these known facts do not support the hypothesis, the preconditions needed for the hypothesis are set up as subgoals. The process for finding a solution is to search from the goal to the initial state, which involves a depth-first search. A trigger is associated with every rule. A rule is selected by the inference engine when the trigger is activated. Only one rule at a time can be activated and the continuation of the selection and firing process is done according to the result of the rule :

• If the result is *unknown*, the rule is put in memory and the rule which gives the value of the first unknown precondition is selected.

• If the result is *known*, the last rule which was put in memory is selected again because the produced result is the value of one of its preconditions.

The process of selection and firing of rules described above is repeated by recursion until the final answer is found; the process can last a long time. In the search for efficiency and performance, unnecessary computations must be avoided. In some cases, there is no need to know the values of all the preconditions of a rule to deduce the value of its consequence. For example, in Boolean logic, if we have the rule :

$$A \text{ AND } B \longrightarrow C.$$

and we know that A is false, then the consequence C is false and there is no need to look for the value of B to conclude that; the set of rules giving the value of B can be pruned.

In systems using fuzzy logic, this avoidance of unnecessary computations is all the more important as computations are more costly in time and memory storage than in systems using Boolean logic. The problem is that little improvement in performance is obtained, if extra computation is avoided only in the case of complete truth (for the operator OR) or of complete falsity (for the operator AND). The solution lies in the setting of thresholds for certain truth and certain falsity. The thresholds for which no further search is required in the execution of the operators are set to 0.8 for certain truth in the operator OR and 0.2 for certain falsity in the operator AND. A rule or fact having a degree of truth larger or equal to 0.8 (resp. less or equal to 0.2) will be considered to be true (resp. false). Thus, the logic takes into account the unknown rules or facts.

1.2 Characteristics of the Predicate Transition Nets Used in the Model

Predicate Transition Nets have been introduced by Genrich and Lautenbach (1981) as an extension of the ordinary Petri Nets (Peterson,

1980; Reisig, 1985) to allow the handling of different classes of tokens. The Predicate Transition Nets used in the model have the following characteristics.

Tokens. Each token traveling through the net has an identity and is considered to be an individual of a given class called variable. Each variable can receive different names. For this model, two classes of tokens are differentiated :

(1) The first class, denoted by P, is the set of the real numbers between 0 and 1, representing the degrees of truth of the facts or items of evidence. The names of the individual tokens of these classes will be p, p_1, p_2.

(2) The second class is denoted by S. The individuals of this class can only take one value. Only one token of this class will travel through the net and will represent the action of the inference engine in triggering the different rules.

Places. Places are entities that can contain tokens before and after the firing of transitions. Three kinds of places are differentiated:

(1) places representing a fact or the result of a rule and containing tokens of the class P or no token at all,

(2) places used by the system as triggers of operators and containing the token of the class S. These places and the connectors connected to these places are represented in bold style in the Figures and constitute the *system net*.

(3) places allowed to contain both kinds of tokens (P and S) and which are used to collect the tokens necessary for the enabling of the transitions of which they are the input places.

The marking of a place is a formal sum of the individual tokens contained in the place. For example, a place A containing a token of the class P, p1, and the token of the class S has the marking M(A) :

$$M(A) = p1 + S$$

Connectors and Labels. Each connector has a label associated with it that indicates the kinds of tokens it can carry. A special grammar is used on the labels to define in what way tokens can be carried. The labels of connectors linking places to transitions contain conditions that must be fulfilled for them to carry the tokens. The labels of connectors linking transitions to places indicate what kind of token will appear in the places after the firing of the tran-

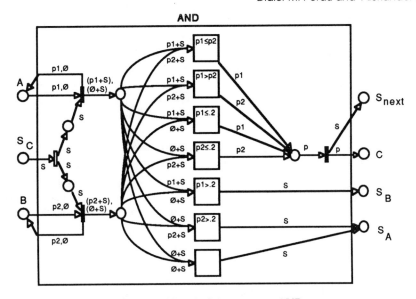

Figure 1. Model of the operator AND.

sition. The following notation in labels is used:

When token names are joined by the sign "+" then the tokens defined by these names must be carried at the same time. For example, the label "p + S" indicates that one token of the class P and one token of the class S have to be carried together at the same time by the connector. When token names are joined by the sign "," then the tokens defined by these names can be carried at different times but not together. For example, the label "p, S" indicates that either a token of the class P or a token of the class S can be carried. Mixing of notation is possible. The label "p+S, S" indicates that the connector can carry either a token of the class P and a token of the class S or only one token of the class S. A connector without label has no constraint on the kind of tokens it can carry.

In some cases, the connector has to carry the token of class S when there is no token of the class P involved in the firing of a transition. The statement "absence of token of the class P" is denoted by the symbol $\emptyset$. This symbol is used in the labels, as if it was a class of tokens, in association with the names of the other classes. The symbol $\emptyset$ is used in the following cases :

(1) The label "S+$\emptyset$" means that the connector can carry a token of the class S, if there is no token of the class P.

(2) The label "(S+p), (S+$\emptyset$)" means that the connector can carry either a token of the class S and a token of the class P, or a token of the

class S, if there is no token of the class P.

Transitions. Transitions have attached to them a predicate, which is a logical formula (or an algorithm) built from the operations and relations on variables and tokens in the labels of the input connectors. The value (true or false) taken by the predicate of a transition depends on the tokens contained in the input places of the transition. When the predicate has the value "true," the transition is enabled and can fire. In the model of the consultant expert system, predicates are conditions on tokens of the class P.

Transitions with predicates are represented graphically with squares or rectangles. The predicate is written inside. Transitions without predicates are represented with bars as in ordinary Petri Nets.

Firing Process. The conditions of enabling of a transition are: (1) the input places contain the combination of tokens specified by the labels of the connectors, and (2) the predicate of the transition is true. If these two conditions are fulfilled, the transition can fire. In the firing process, tokens specified by the input connectors are withdrawn from the corresponding input places and tokens specified by the output connectors are put in the output places.

1.3 Logical Operator Models

In order to construct the model of the expert system using Predicate Transition Nets, it is necessary to first construct models of the logical operators AND, OR, and NOT. The results are shown in Figures 1, 2 and 3. Let us describe now what happens in the operator AND (the operators OR and NOT behave in a similar way). The operator drawn in Figure 1 realizes the operation :

$$A \ \text{AND} \ B \longrightarrow C.$$

It can be represented as a black box, having three inputs: A, B, and S_C (the trigger) and six outputs : C (the result), A, B (memorizing of the input value) and three system places S_A, S_B, and S_{next}. Only one of those system places (represented in bold style in the figures) can have a system token at the output. S_{next} will contain a system token, if the result of the operation is known (i.e., if C contains a token of the class P). This shows that the next operation can be performed. If the result is unknown, that is, the two inputs are not suffi-

cient to yield a result, the system token is assigned to S_A or S_B in order to get the values of these unknown inputs. A system token will be assigned to S_A if (i) C is unknown and (ii) A is unknown or if A and B are both unknown. The system token will be assigned to S_B if C is unknown and only B is unknown.

The execution of the operation will start only if there is a system token in S_C. We denote by S_C the trigger place of the operator computing C. As soon as there is a token in S_c, the two input transitions are triggered by the allocation of a system token (S) at the input places of these transitions. The values of A and B are therefore reproduced in A and B and in the output place of each of the transitions. These places contain also a system token, that will ensure the enabling of the following transition (i.e., that the two inputs are present). These two places are the input places of seven different transitions which have disjoint conditions of

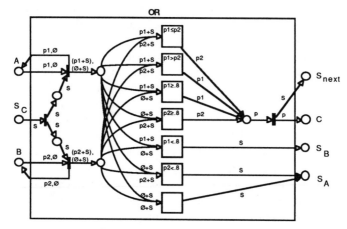

Figure 2. Model of the operator OR.

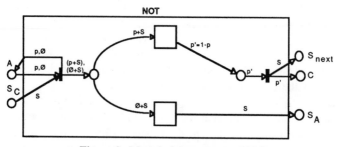

Figure 3. Model of the operator NOT.

enabling. Only one of these transitions can be enabled and can fire. At the firing, the result, if any, is given in the result place and then in C, while the system token is assigned either to S_{next}, or to S_A, or to S_B.

These operators can be compounded in super-transitions. The model can be generalized to operators with more than two inputs by combining these basic operators.

An example of the use of these logical operators is shown in the next section, where a simple inference net is modeled and the search process in this net is simulated.

1.4 Dynamic Representation of the Knowledge Base

The connection of the super-transitions representing the logic operators to places representing the items of evidence leads to a dynamic representation of the knowledge base. It allows to show explicitly how the inference engine scans the knowledge base. By running a simulation program, we can see in real time what the steps of reasoning are, the possible deadlocks, or mistakes. It allows one to identify the parts of the knowledge base where the knowledge representation is incorrect.

Let us consider the simple symbolic system containing the following rules :

$$\text{if A and B} \longrightarrow \text{E}$$
$$\text{if C and D} \longrightarrow \text{F}$$
$$\text{if E or F} \longrightarrow \text{G}$$

The representation of this simple symbolic system using the Predicate Transition Net models of the logic operators is shown in Figure 4. The interface module with the user has been added through the places IA, IB, IC and ID, where the user can enter the degrees of truth of A, B, C, and D.

The simulation of the propagation of the tokens in this net allows one to observe the reasoning process followed by the system. The mapping of the different places of the net at each step of the process of the simulation is shown on Table 1.

The search for the degree of truth of the goal G starts when the system token, S, is put in the system place S_G, at the beginning of the search (step 1). The degree of truth of G cannot be evaluated when the operator OR is executed. The system token is therefore assigned to S_E to check the subgoal E (step 2). The execution of the operator AND cannot lead to a result for E and the system token is allocated to S_A (step 3), which triggers an interactive session with the user to get the degree of truth of A. The user enters this value (say, 0.9) through place IA (step 4) which is assigned to A, while the system token is assigned to S_E (step 5). Since, the degree of truth of A is larger than 0.2, the result of the operator AND cannot be given in E and the system token is assigned to S_B (step 6) to get the degree of truth of B (say, 0.8) through IB (step 7). The system token is then reassigned to S_E to trigger the operator AND (step 8), which can now be executed. The minimum of the degrees of truth of A and B, 0.8, is placed in E, while the system token is assigned to S_G (step 9). Since the degree of truth of E is equal to 0.8, the operation OR can be per-

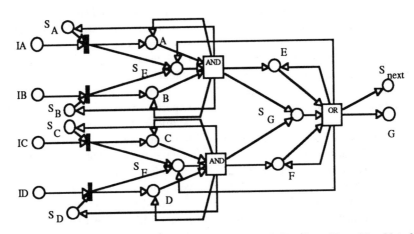

Figure 4. Representation of a simple symbolic system using the Predicate Transition Nets formalism.

	A	IA	B	IB	C	IC	D	ID	E	F	G	S_A	S_B	S_C	S_D	S_E	S_F	S_G	S_{next}
Step 1																		S	
Step 2														S					
Step 3											S								
Step 4		0.9									S								
Step 5	0.9														S				
Step 6	0.9													S					
Step 7	0.9			0.8										S					
Step 8	0.9		0.8												S				
Step 9	0.9		0.8						0.8								S		
Step 10	0.9		0.8						0.8	0.8									S

Table 1. Mapping of the Places at the different steps of the simulation.

formed to produce the result G equal to 0.8. The system token is allocated in S_{next} (step 10). The subgoal F has not been checked and the part of the net which is used to evaluate F has been pruned.

2.0 Use of the Expert System for the Fusion of Inconsistent Information

An important problem faced by decision making organizations is the inconsistency of information, which can degrade substantially their performance. This inconsistency can be attributed to different causes: inaccuracy in measured data, lack of sensor coverage, presence of noise, bad interpretation of data. In a military context, inconsistency of information can also be explained by the enemy's attempt to mislead the organization about his actions through the use of decoys or jamming techniques. This presence of inconsistent information jeopardizes the successful execution of an organization's mission. Three strategies to fuse inconsistent information are considered in this paper : (1) ignore information sharing, (2) weighted choice among contradictory sets of data, and (3) use of an expert system that has additional knowledge on the problem to be solved.

The first strategy occurs when the decision maker performing the information fusion uses only his own assessment and ignores the assessment of the other decision maker. This strategy is related to the way a human being assigns value to information that is transmitted to him. The study of Bushnell, et al. (1988) develops a normative-descriptive approach to quantify the processes of weighting and combining information from distributed sources under uncertainty. Their experimentation has shown that one of the human cognitive biases, which appears in the execution of a task, is the undervaluing of the communications from others, and is independent of the quality of the information received. The decision maker is, therefore, expected to have the tendency to overestimate his own assessment and to assign a lower value to the others' assessments.

The second strategy is to perform a weighted choice among the contradictory assessments that are transmitted to him and compared to his own. This weighting strategy involves the confidence that can be given to the information, which depends on the manner in which this information has been obtained, or on its certainty. In many models of organizations facing this problem of inconsistent information and using the weighted choice strategy, measures of certainty are the basis for the weighting of different items of evidence. Among the methods used, the Bayesian combination has given valuable results.

The third strategy involves the use of an expert system. Expert systems can consider additional knowledge and facts that would be too costly in terms of time, effort, and memory storage to be handled efficiently by the decision maker on his own. For each instance of contradictory data, it can check if their values are consistent with the knowledge it has and give an indication of their correctness. With this additional attribute, the decision maker can

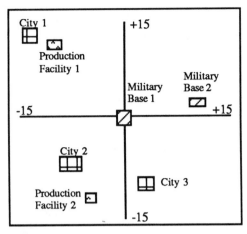

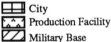

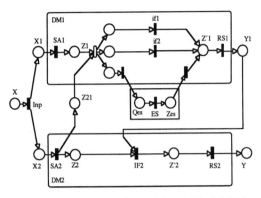

Figure 5. Location of facilities to be defended
by the organization.

Figure 6. Petri net of the hierarchical 2-DM
organization.

perform a more precise information fusion.

To illustrate how these strategies modify the measures of performance of an organization and to emphasize the role of an expert system in the fusion of inconsistent information, an illustrative application will be used.

2.1 Command and Control in an Air Defense Problem

Mission and Organization: The illustrative application involves an organization, whose mission is to defend a set of facilities against attacking missiles. This set of facilities consists of three cities, two military bases and two production facilities located in a square 30 miles wide, as shown in Figure 5. To destroy incoming missiles, the organization can either use a laser beam or send an antimissile rocket. The laser beam is used in case of urgency, when the time before the missile hits its target is less than a certain threshold. The antimissile rocket is used when enough time is available. Both weapons require different targeting solutions. The performance of the organization is measured by its ability to send the right weapon at the right place for each incoming threat.

The considered organization is a hierarchical two-decision makers organization with the

Petri net representation (Remy et al., 1987) shown in Figure 6. The two decision makers, DM1 and DM2, perform their own situation assessment producing the results Z1 and Z2. DM2 sends Z21, which is equal to Z2, to DM1 who is in charge of performing the information fusion with one of the three strategies available. One of them is to use an expert system. Using the revised situation assessment Z'1, the response Y1 is selected and transmitted to DM2. DM2 takes into account this new information in his information fusion stage and realizes the final response selection of the organization, Y.

Inputs and situation assessments: Each decision maker receives as input two points of the trajectory of the missile. The first one is its position at time t, which is the same for the two decision makers to make sure they are assessing the same missile. The second point is determined by the tracking center of each decision maker. The tracking center is defined as the sum of the human and hardware means assembled to process the information. The use of decoys by the enemy and the presence of noise result in the second data points not being the same for each of the decision makers. When this is the case, we assume that one of the two is the actual one. In addition to these different coordinates, the input also contains the confidence factors associated with each position. These confidence factors have been generated by a preprocessor (say, a tracking algorithm) and measure the quality that can be attributed to each set of data.

After receiving these inputs, the two decision makers, DM1 and DM2, perform the same situation assessment. DM1 (resp. DM2) computes the velocity of the missile and evaluates its impact point, according to the set of coordinates he has received, and produces the result Z1 (resp. Z2).

Information fusion of DM1: In his information fusion stage, DM1 first makes the comparison between Z1 and Z21. If they are equal, Z'1 = Z1 is produced. If they are different, DM1 must resolve the matter by choosing from the three different strategies described in the previous section.

The first one is to ignore information sharing. In this case, DM1 produces Z'1 = Z1 without considering the situation assessment Z21, transmitted to him by DM2.

The second strategy is the weighting of the information according to the confidence factors associated with each set of data. DM1 considers the confidence factors Conf1 and Conf2 given with the input and which measure the quality of the information to choose Z1 or Z21. If Conf1 is greater than or equal to Conf2, DM1 produces Z'1 = Z1. In the opposite case, DM1 produces Z'1 = Z21.

The last strategy involves the use of an expert system. The simple knowledge base system, which has been developed for this application, evaluates the degree of threat a missile represents as a function of the distance between the location of the different facilities and its impact point estimated by the user. A more sophisticated system could make the assessment of the threat by taking into account the type of missile, the geographical aspect of the area, the direction of winds, and the value to the enemy of the threatened facility. The threat assessment of the missile is done for the two possible trajectories, one after another. If the first threat assessment shows that the target is one of the facilities with enough certainty, the computer stops its search. In the opposite case, the computer also evaluates the threat that the missile would pose, if it were to follow the second trajectory. The answer of the expert system consists of two numbers between 0 and 1 representing the severity of the threat posed by the missile (according to each assessment). When the answer is given, DM1 does not use a strategy in which a comparison is made with a result from an internal algorithm, as in Weingaertner and Levis (1989). This is due to the fact that the decision maker does not have enough data on his own to be able to double check the answer of the decision aid. If the degree of threat according to the assessment of DM1 is greater than or equal to the one according to the assessment of DM2, the result is Z'1 = Z1. In the opposite case, the result is Z'1 = Z21.

Response of the organization: Having chosen the trajectory that seems to be the most likely, DM1, in his response selection stage, determines the type of threat the missile represents by computing the time before impact and sends it to DM2 with the fused information. DM2, in his information fusion stage selects the weapon to use and performs the targeting solution in his response selection stage.

2.2 Measures of Performance

The measures of performance considered in this paper are workload (Boettcher and Levis, 1982; Levis, 1984), timeliness (Cothier and Levis, 1986) and accuracy (Andreadakis and Levis, 1987). They have been defined for the two possible types of interaction between the computer and the user:

The user initiated mode in which the decision maker enters all the data he has in a specified order and the machine produces a result. Not all entered data may be needed by the machine in its search process.

The computer initiated mode in which the user enters specific data only in response to requests from the computer.

Thirty-three equiprobable inputs to the organization have been considered. Twenty-four inputs contain inconsistent information. We assume that for half of these inconsistent inputs, the tracking center of DM1 is correct (the tracking center of DM2 is correct for the other half because we assume that for each input, one of the two contradictory positions is correct).

2.2.1 Workload

The evaluation and the analysis of workload in a decision making organization uses an information theoretical framework (Levis, 1984). The activity of a decision maker is evaluated by relating, in a quantitative manner, the uncertainty in the tasks to be performed with the amount of information that must be processed to obtain certain results. The information theoretic surrogate for the cognitive workload of a decision maker is computed by adding all the entropies of all the variables used to model the procedures he uses to perform his task.

2.2.2 Timeliness

The measure of timeliness considered in this application is related to the response time of the organization. A deterministic processing time has been associated with every algorithm. Each processing time could have been described by a probability density function and the probability density function of the response time could have been computed (see Andreadakis and Levis, 1987). The use of a stochastic model does not add to the example, but it would be the model to use in an experimental investigation. For the strategy involving the use of the expert system, the time to give an answer has been computed using the expert system model described in the first section of the paper. The response time of the expert system is a function of the number of rules scanned by the system for each input to the organization and of the number of interactions with the user. This time is likely to vary with the mode of interactions used.

We assume that DM1 and DM2 perform their situation assessment concurrently and synchronously, and that the same amount of time is needed by the two to give an answer. Therefore, only one of the two processing times is considered. T_{SA1}, (resp. T_{RS1}, T_{SA2}, T_{IF2} and T_{RS2}) denotes the time needed to execute DM1's situation assessment algorithm (resp. DM1's response selection, DM2's situation assessment, information fusion and response selection). $T_{IF1(i)}$ is the time needed to perform the information fusion using a pure strategy i [i = 1, 2, 3]. $T_{IF1(3)}$ is a function of the average response time of the expert system computed from its response time for all the inputs. The response time for the strategy i, $T(i)$, is therefore:

$$T(i) = T_{SA1} + T_{IF1}^{(i)} + T_{RS1} + T_{IF2} + T_{RS2}.$$

The response time for each mixed strategy (p_1, p_2, p_3) is given by a convex weighting of the response time for each pure strategy. If $T(p_1, p_2, p_3)$ denotes the response time of the organization when the strategy (p_1, p_2, p_3) is used, we have :

$$T(p_1, p_2, p_3) = p_1 T(1) + p_2 T(2) + p_3 T(3)$$

2.2.3 Accuracy

Accuracy of the organization has been evaluated by comparing the actual response of the organization with the desired or optimal response expected for each input. This desired response is known to the designer. A cost of one has been attributed when the incorrect type of weapon is used or when the target point is not accurate. For each input X_j having a probability $p(X_j)$, the use of the pure strategy i generates the response Y_{ij} which is compared to the desired response Y_{dj}. The cost function $C(Y_{ij}, Y_{dj})$ has the following characteristics:

$$C(Y_{ij}, Y_{dj}) = \begin{cases} 1 & \text{if } Y_{ij} \neq Y_{dj} \\ 0 & \text{if } Y_{ij} = Y_{dj} \end{cases}$$

The accuracy $J(i)$ obtained for the pure strategy i is:

$$J(i) = \sum_j p(X_j) C(Y_{ij}, Y_{dj})$$

The accuracy for the mixed strategy (p_1, p_2, p_3), $J(p_1, p_2, p_3)$, is obtained by computing the convex combination of the accuracy for each pure strategy:

$$J(p_1, p_2, p_3) = p_1 J(1) + p_2 J(2) + p_3 J(3)$$

Consequently, J represents the probability that an incorrect response will be generated. The lower the value of J, the better the performance is. The next section provides an analysis of the results obtained by using these measures of performance.

3.0 Results and Interpretation

Using the method described above, measures of performance have been evaluated for the three strategies. For the strategy involving the use of an expert system, we have considered two different options for dealing with uncertainty in the firing of rules, fuzzy logic or Boolean logic; and two modes of interaction between the user and the decision aid: user ini-

	Strategy 1 Ignoring other assessment	Strategy 2 Weighted Choice	Expert System Fuzzy Logic		Expert System Boolean Logic	
			user initiated	computer initiated	user initiated	computer initiated
J cost	0.36	0.27	0.21	0.21	0.24	0.24
T seconds	18.24	18.96	21.02	20.85	20.97	20.36
G1 bits/symbol	63.41	64.92	70.29	70.29	65.97	65.97
G2 bits/symbol	43.92	43.85	43.24	43.24	43.29	43.29

Table 2. Measures of performance for the three strategies.

tiated mode or computer initiated mode. The results are summarized in Table 2.

3.1 Measures of Performance

3.1.1 Pure Strategies.

The three first columns of Table 2 display the measures of performance (MOPs) of the organization for each pure strategy. These results show that taking more knowledge into account, either about the way data are obtained, in the case of the weighted choice strategy, or about the meaning of information, when the expert system is used, yields greater accuracy. Accuracy is an important issue for the kind of mission this type of organization is expected to carry out. The results show also that taking into account more knowledge requires handling more data. Therefore, more time and more effort are required. This increased workload is caused more by the extra decisions that must be made when the knowledge is taken into account, than by operations or manipulation done with the additional knowledge. These manipulations are done by the decision aids, out of DM1's control.

When DM1 ignores the situation assessment of DM2, very few operations are performed. The response time is the smallest of the three. If the measure of timeliness is the ability of the organization to give a response as quickly as possible, this strategy leads to a more timely response than the other two. The simplicity of the algorithm results in low workload for DM1 in comparison with the other strategies, because DM1 handles fewer variables. This strategy has low accuracy in comparison with the other strategies, because the choice made on the information to be fused is arbitrary and has no rational justification. Thus, a clear assessment of the cost and value of coordination can be made.

For the weighted choice strategy, no operation on variables received is performed. DM1 makes only a comparison between the weights of the information. We have assumed that the weighting process was carried out outside the organization by a preprocessor and, consequently, DM1 performs only a few operations more than in the first strategy. Therefore, workload and response time are slightly larger than for the first strategy because of the extra information obtained by comparing the confidence factors. An increase of 3.9 percent in response time and of 2.4 percent in the workload of DM1 is found. The measure of how the data have been obtained, given through the confidence factors, brings a large gain in accuracy. An improvement of 25 percent in the accuracy of the organization in comparison to the first strategy is observed. These results show, as expected, that taking into account the quality of information plays an important role in the accuracy of the organization, without substantially degrading the other measures of performance.

When the expert system is used, the increase in workload of DM1 is about 8.3 percent from the level of strategy 2, and 10.8 percent from the level of the first strategy. This can be explained by the handling by DM1 of the assessments given by the expert system. These assessments are variables which have greater entropies and which require more processing.

The increase in response time (of 10.8 percent from the level of strategy 2 and of 15.2 percent from the level of strategy 1) is mainly caused by the time taken by DM1 to interact with the system and the time needed to get the answer. This response time of the expert system can get larger as the size of the knowledge base and of the magnitude of the problem to solve increase. In the example, the simplicity of the expert system hides the real effect on timeliness, which can be expected with the use of such an interacting system. The gain in accuracy is very significant, about 22 percent, in comparison with the accuracy reached with the second strategy and 41.7 percent from the level reached when the situation assessment of the other DM is ignored. This shows the extent to which the accuracy is improved when additional knowledge is used to verify the correctness of information. By using the expert system to evaluate the threat and to estimate the severity of the threat for each possible trajectory, DM1 has a broader assessment which allows him to perform more accurate information fusion.

Finally, we note that the workload of DM2 remains almost constant for all the strategies. A variation of 1.5 percent can be observed. He always uses the same algorithms, and only the different distributions of the variables of the algorithms obtained, when different strategies are used by DM1, explain this small variation in his workload.

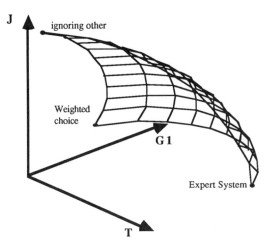

Figure 7. Locus of the measures of performance attained by the organization.

3.1.2 Mixed Strategies

The performance measures (accuracy, timeliness, and workload of DM1) reached by the organization, when mixed strategies are used by DM1 in his information fusion stage, have been obtained using the CAESAR (Computer Aided Evaluation of System Architectures) software. Measures of performance have been evaluated for all mixed strategies and have led to a surface in the space (J-T-G1) represented in Figure 7. The projections of this surface on the Accuracy-Workload (J-G1), and Timeliness-Workload (T-G1) planes are drawn in Figure 8. Measures of performance reached for each pure strategy are located at the three cusps of the figures. The convex combination of any two pure strategies gives a U-shaped curve (Boettcher and Levis, 1982), which can be explained by the fact that when a mixed strategy is used, there is an additional activity due to switching from one algorithm to another.

The projection of the surface of the measures of performance on the accuracy-timeliness plane (J-T) gives the triangle shown in Figure 9, which shows the performance attained by the organization. The corners of this triangle indicate the level reached in accuracy and response time for each pure strategy. For all binary variations between pure strategies or for all successive binary combinations of mixed strategies, J and T are linear combinations of each other. Figure 9 shows clearly the trade-off between response time and accuracy and how the requirements of the mission will justify a strategy. Thus, if the requirements in accuracy are too binding, the strategy of ignoring information sharing will not be acceptable. In the same way, if the time available to process each input is too short, the expert system would be useless because too much time will be needed to perform the information fusion.

3.2 Effect of the Mode of Interaction

The effect of the mode of interaction on the measures of performance is shown in the last four columns of Table 2. There is no change in accuracy or workload; however, a slight change in timeliness is observed. This is because, in the user initiated mode of interaction, all the data which have a chance to be processed by the expert system are entered at the beginning of the session. In the example, the position of the impact points according to the two different

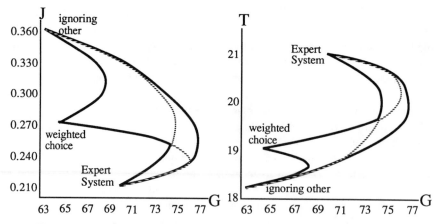

Figure 8. Mixed strategies-accuracy/timeliness versus workload for DM1.

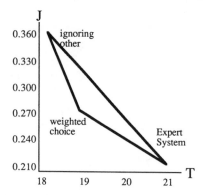

Figure 9. Mixed strategies: accuracy and timeliness of the organization.

situation assessments are entered, even if the first set is sufficient to assess the threat. Therefore, more time is needed than in the computer initiated mode where data are entered at the request of the system during the search.

It is important to note that in the air defense, no workload have been assigned to the process of entering the information in the expert system. The process consists only of replication of the information the decision maker already has. If the inputs asked by the expert system do not correspond to the data the decision maker has, he would have to perform some operations to deduce these inputs from the information. Let us consider an example where the decision maker has computed, or received from another member of the organization, the value of the speed of an object being analyzed. If the expert system asks the decision maker the question: "speed of the object : [possible answer: low, moderate, high]," the

decision maker will have to deduce from the actual value of the speed the attribute asked by the system. A small algorithm will have to be executed, increasing his workload. Therefore, it can be expected that, in this case, a change in workload similar to the change in response time would be observed. This issue raises the problem of the adequate design of the expert system, or more generally, of the decision aid. The mode of interaction has to be thought through very carefully to avoid an unnecessary increase in the workload of the decision maker and in the response time.

3.3 Fuzzy Logic versus Boolean Logic

For this illustrative application, the levels of performance reached when different expert systems are used have been studied. The performance achieved with an expert system using fuzzy logic as the means of inference, which has been developed for the example, has been compared to the performance obtained by using an expert system that does not deal with uncertainty and uses Boolean logic. This version of the expert system has been obtained by changing the mapping functions (only values 0 and 1 could be processed instead of the real numbers between 0 and 1). It has been assumed that a statement having a degree of truth greater (resp. smaller) than 0.6 was true (resp. false). Therefore, the assessment of the threat of the missile for each trajectory has only the values true or false. The different measures of performance obtained for the two systems are summarized in the last four columns of Table 2.

The organization has a response time slightly lower with an expert system using Boolean logic than with the expert system using fuzzy logic (2.3 percent). This is due to the fact that by assigning the value true or false to the severity of threat, the system can reach a conclusion (which is not always the best one) by examining fewer possibilities. It can prune a larger part of the knowledge base than the fuzzy logic system when it reaches the conclusion that the missile is threatening a specific facility. When this conclusion is reached for the first possible trajectory, the other trajectory is not examined. This results in a shorter time to produce the answer and in fewer interactions with the user and therefore a shorter response time.

Since the expert system with Boolean logic assesses the threat only with the value true or false, the answer of the expert system has a lower entropy. The workload of the decision maker is therefore lower (about 6.8 percent) when he uses the expert system with Boolean logic than when he uses the expert system with fuzzy logic.

By pruning a larger part of the knowledge base when it reaches a conclusion, the system has more chance to make the wrong assessment of the threat. The results show that, indeed, the system with Boolean logic exhibits lower accuracy than the system with fuzzy logic. The level of accuracy is, nevertheless, better than for the two other strategies expected to be used in the information fusion stage and is explained by the fact that more knowledge is taken into account in the information fusion process.

4.0 Conclusion

In this paper, a model with fuzzy logic as a means for dealing with uncertainty has been developed using the Predicate Transition Net formalism. A method to make time-related measures from this representation has been introduced, taking into account the portion of the rule base scanned by the system and the number of interactions. Then, the assessment of the role of an expert system has been made through the study of an example which involves a two decision maker organization facing the problem of fusion of inconsistent information. The decision makers must identify the trajectories of threats that they then have to destroy to protect a set of facilities. In the example, the expert system helps the decision maker to clarify the contradictory situation assessment he has to fuse. This strategy has been compared to two others expected to be used in this situation: (1) ignoring the assessment of the other decision maker, (2) making a weighted choice among the two contradictory situation assessments, by taking into consideration the way the data used to produce these assessments have been obtained by each decision maker. Measures of performance (workload, timeliness, and accuracy) have been evaluated. The results show that the use of the expert system improves significantly the accuracy of the organization, but requires more time and increases the workload of the decision maker using it. The comparison of the two modes of interaction between the user and the system has shown variations in workload and in response time: the computer initiated mode requires less workload and less response time for the same level of accuracy. This result tends to show that the design of an interacting decision aid must take into account not only the characteristics of the problem to be solved, but also the way the decision maker would use it.

References

Andreadakis, S. K. and Levis, A. H., 1987, "Accuracy and Timeliness in Decisionmaking Organizations," *Proc. 10th IFAC World Congress*, Pergamon Press, New York.

Boettcher, K. L. and Levis, A. H., 1982, "On Modeling Teams of Interacting Decisionmakers with Bounded Rationality," *Automatica*, vol. 19, No.6 (1983), pp. 703-709.

Bushnell, L. G., Serfaty, D. and Kleinmann, D. L., 1988, "Team Information Processing: A Normative-

Descriptive Approach," *Science of Command and Control: Coping with Uncertainty,* Edited by S. E. Johnson and A. H. Levis, AFCEA International Press, Fairfax, Virginia.

Conant, R. C., 1976, "Laws of Information Which Govern Systems," *IEEE Transactions on Systems, Man and Cybernetics,* vol. SMC-6, No.4, pp. 240-255.

Cothier, P. H. and Levis, A. H., "Timeliness and Measures of Effectiveness in Command and Control," *IEEE Transactions on Systems. Man and Cybernetics,* SMC-16, no 6.

Genrich, H. J., Lautenbach, K., 1981, "System Modeling with High Level Petri Nets," *Theoretical Computer Science,* 13 , pp. 109-136.

Giordona, A. and Saitta, L., 1985, "Modeling Production Rules by Means of Predicate Transition Networks," Information Sciences, 35, pp. 1-41.

Johnson, L. and Keravnov, E. T., 1985, *Expert Systems Technology, A Guide,* Abacus Press, Tunbridge Wells.

Levis, A. H., 1984, "Information Processing and Decisionmaking Organizations: A Mathematical Description," *Large Scale Systems* , vol. 7, pp. 151-163.

Peterson, J. L., 1980, *Petri Net Theory and the Modeling of Systems,* Prentice Hall, Inc., Englewood Cliffs, New Jersey.

Reisig, W., 1985, *Petri Nets, An Introduction,* Springer Verlag, Berlin.

Remy, P., Levis, A. H., and Jin, V. Y. Y., 1988, "On the Design of Distributed Organizational Structures," *Automatica,* vol. 24, no.1, pp 81-86.

Weingaertner, S.T. and Levis, A. H., 1989, "Evaluation of Decision Aiding in Submarine Emergency Decisionmaking," *Automatica,* vol. 25, no.3.

Whalen, T. and Schott, B., 1983, "Issues in Fuzzy Production Systems," *International Journal of Man-Machine Studies,* vol. 19, pp. 57-71.

Zadeh, L. A., 1965, "Fuzzy Sets," *Information and Control,* vol. 8, pp. 338-353.

Zadeh, L. A., 1983, "The Role of Fuzzy Logic in the Management of Uncertainty in Expert Systems," *Fuzzy Sets and Systems,* vol. 11, pp. 199-227.

---

* This work was conducted at the MIT Laboratory for Information and Decision Systems with support provided by the Basic Research Group of the Technical Panel on C3 of the Joint Directors of Laboratories through the Office of Naval Research under contract no. N00014-85-K-0782.

9

Planning with Uncertain and Conflicting Information

Daniel Serfaty, Elliot E. Entin, and Robert R. Tenney

Introduction

Several paradigms to describe command and control (C²) processes have been introduced in the past 10 years. These include Lawson's Process Model [1, 2], Wohl's SHOR (Situation, Hypothesis, Options, Response) paradigm [3, 4], and the HEAT (Headquarters Effectiveness Assessment Tool) Adaptive Control paradigm [5]. The HEAT paradigm is shown in Figure 1. All three models view the decision making process as a cycle comprising an assessment of the situation, a comparison of the current situation to the situation desired by the decision makers, and a selection of an option (or options) to maintain or achieve the desired situation.

An outstanding issue for these models is whether the planning cycle is best viewed as an evolutionary process, supporting the "rolling plan" concept suggested by Cushman [6], or as a sequence of independent processes. If planning cycles are independent, then each cycle can be understood (and analyzed) without regard to those that came before or after. This point of view does not imply that decision makers have no memory of the past, only that they do not actively reach decisions at one cycle to foster the decision making process at a later cycle.

Conversely, the "rolling plan" point of view implies that the understanding of the situation assessment process requires consideration of potential future perceptions and decisions. It also implies that understanding the option selection process requires a consideration of plans and decisions that may be made in the future. This point of view stresses learning in order to reduce uncertainty as an important characteristic of decision making and planning. Lawson [2] and Wohl [7] propose *rate of change* of uncertainty as an indicator of effectiveness for C² systems. Indeed, probing and hedging strategies that allow decision makers to learn more about their environment before selecting a course of action can be understood only from the perspective of decision making and planning as evolutionary processes.

Experiment Hypotheses

The purpose of this study was to better understand the planning process conducted by a military headquarters. The study tested three major hypotheses:

Hypothesis 1

Our first hypothesis is that measures taken of the planning process are correlated over several

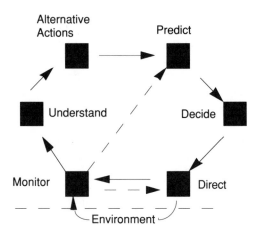

Figure 1. HEAT Adaptive Control paradigm for C².

planning cycles. The arguments for this hypothesis are found in normative theories for estimation and information seeking, as well as in normative-descriptive decision making theories from cognitive science. Estimation theory tells us that optimal estimators (e.g., Bayesian estimators, Kalman filters) combine prior estimates of the random variable under consideration with new measurements to yield posterior estimates. Normative-descriptive theories of hypothesis generation and selection from cognitive science tell us that people use prior hypotheses to adjust their beliefs in response to new data.

Normative theories of information seeking that have been developed for the control of stochastic systems typically involve parameters that are only partially known to the controller or decision maker; this is known in military terms as the "fog of war." The control of these systems typically involves an element of *active* learning wherein information-seeking strategies (e.g., probing) are selected by decision makers in anticipation that they will better understand their system and use this understanding to advantage in *future* decisions.

Hypothesis 2

Our second hypothesis is that process measures, such as the strength of the commanders' belief in their assessments, are correlated to the commanders' confidence in the information they receive. Arguments for this hypothesis are found in Estimation theory and in Stochastic Optimal Control theory. Estimation theory tells us that information should be given a weight that is inversely proportional to the confidence or certainty that can be attributed to that information. Stochastic Optimal Control theory tells us that the optimal control of a stochastic system is achieved through a control policy that is conditioned on the uncertainty surrounding the state of the system at any time [8]. Our argument does not presume that humans are optimal estimators or optimal controllers, but that they are sufficiently rational to exhibit important characteristics found in optimal estimators and optimal controllers.

Hypothesis 3

Our third hypothesis is that decision makers are sensitive to the order of presentation of information that confirms or disconfirms prior beliefs when they assess a situation. The argument for this hypothesis is based on several models that have been proposed in cognitive science to describe human decision making.

It is generally acknowledged that decision makers do not learn nearly as much from newly acquired information as the information warrants [9-11]. Tolcott et al. [12] note that decision makers usually place undue confidence in the correctness of prior decisions and assessments, supporting the idea that people do not readily learn from experience. Decision makers tend to anchor to prior beliefs by inadequately adjusting their position upon consideration of new information.

Einhorn and Hogarth [13,14] offer a different view of the picture. Their research indicates that people tend to place greater emphasis on newer information than on prior information. Einhorn and Hogarth have developed a number of mathematical models of situation assessment and decision making based on these principles. These models capture both the anchoring and adjustment process, as well as the process decision makers employ for dealing with confirming and disconfirming evidence. Section V will focus on the validation of these models using the experimental data collected in this study.

Experimental Method

Information integration and planning processes were studied in the context of a table-top exercise. The protocol is an elaboration of two pencil-and-paper games, one employed by Entin et al. [15] and one described by Tolcott et al. [12]. As a background to the exercise, commanders were told that the relations between the Soviet Union and the United States have greatly deteriorated and that a clash of forces in several parts of the world is imminent. The subjects were asked to assume command of the 52nd Mechanized Infantry Division, stationed either in the Fulda Gap, West Germany, or near the demilitarized zone (DMZ) in Korea. Their primary duty was to consider all the available intelligence information and predict the enemy's most likely main avenue of attack. Based on this estimate of enemy intent, they were to construct a plan allocating and reassigning resources to best thwart the attack.

Each trial in the experiment had four phases:

• First, the commander received a background intelligence briefing describing the intelligence analysts' best estimate of the enemy's avenue of attack, and was asked to give his own estimate of the situation and his confidence in the information provided.

• Second, the commander was shown a map of the deployment of his own troops and the best estimate, based on more recent intelligence, of the enemy's positions and strengths. He was asked to give a second estimate of the enemy's most likely attack approach and his confidence level. The commander then developed a plan adjusting troop positions and deploying reserve units.

• Third, the commander received the first of two intelligence updates on the Orange force dispositions. Given this new information, the commander had an opportunity to revise his attack approach estimate, confidence, and plan.

• Fourth, the commander was given a second intelligence update on the Orange force dispositions. The commander was then allowed one last time to revise his attack estimate, confidence, and plan.

One captain and two majors from the Army's 101st Airborne Division, Fort Campbell, Kentucky, served as subjects. All the officers had at least 12 years of service and had commanded at the platoon level; the two majors had served as company commanders.

The experiment manipulated three independent variables:

• *Enemy Avenue of Attack* — defined as the intelligence analysts' prediction of the enemy's main avenue of attack.

• *Information Source Reliability* — defined as the confidence or reliability (low or high) that the intelligence analysts (purportedly) attributed to the source of the enemy deployment information provided to the subjects.

• *Sequence* — After reviewing the initial intelligence briefing, each subject received a sequence of three additional messages. The information in these messages either confirmed or disconfirmed the information regarding the enemy's intended avenue of attack given in the initial briefing. Sequence was thus defined as the order in which confirmatory and contradictory evidence was presented in the three messages, as shown in Table 1.

Three dependent variables were assessed

Sequence	Situation Map	Update One	Update Two
1	Confirm	Confirm	Disconfirm
2	Confirm	Disconfirm	Disconfirm
3	Disconfirm	Confirm	Confirm
4	Disconfirm	Disconfirm	Confirm

Table 1. Sequences of incoming evidence.

after the initial intelligence briefing and after each of the three messages for each trial:

• subject's prediction of the avenue of the enemy's attack,

• subject's estimation of the probability that his predicted attack approach would be used (i.e., the strength of his belief in his own hypothesis); and

• subject's confidence in the information received at that point.

Four additional dependent measures were derived from the initial plans developed by the commanders and their subsequent revisions:

• the number of units given orders to move to a new position;

• the number of units given orders reassigning them to a new authority;

• the number of orders to move or reassign units in the future (including conditional orders); and

• the number of units reassigned or moved to a Division or to a central position. This measure was used to assess the Blue force's hedging tendencies.

Experimental Results

This section describes the results obtained in the experiment, interpreted to identify behavior trends and to test our three hypotheses. The decision process in the experiment consists of two separate but interacting processes. First is situation assessment, in which the commanders combine, in a sequential manner, incoming information or evidence (e.g., map) with previously held beliefs to produce an updated hypothesis about the likely enemy avenue of attack. Second is sequential planning, in which the commander uses knowledge of his troops' positions and his current beliefs to make decisions concerning the resulting *hypothetical* movement of his troops to best defend against the possible enemy attack. Note that no actual movement of troops occurs; this is on-paper planning only. Thus, these decisions are made in a non-reactive hostile environment, and we assume a "separation principle" between the two processes. The dual process of situation assessment and planning corresponds to the "UNDERSTAND" and "DECIDE" nodes in the HEAT hexagon paradigm of Figure 1.

Situation Assessment Results

The following were the key results regarding our three hypotheses during situation assessment:

Observation 1: The sequence in which the new evidence is presented makes a difference. This supports hypothesis 3.

Figures 2 through 5 show the probabilities given by commanders (their strength of belief in their own hypotheses about the avenue of attack) and their confidence ratings for the information presented to them for each of the four sequences of incoming evidence (See Table 1). To simplify the data presentation, all probabilities refer to the initial hypothesis. The order effect is omnipresent in the data. For instance, compare sequences CDD in Figure 3 and DDC in Figure 5. Each contains two pieces of disconfirmatory evidence and one piece of confirmatory evidence. However the final average probabilities (strength of belief) are drastically different. These numbers not only reflect a quantitative difference but also represent an opposite final hypothesis as to the likely avenue of enemy attack. This result shows clear evidence of the phenomenon of

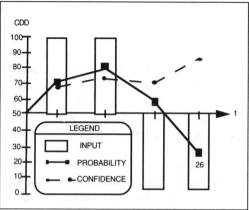

Figure 3. Situation assessment results (probability and confidence) for sequence CDD.

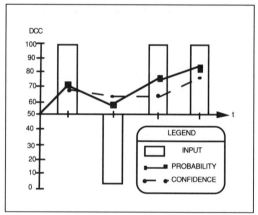

Figure 4. Situation assessment results (probability and confidence) for sequence DCC.

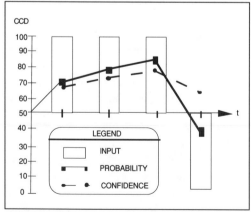

Figure 2. Situation assessment results (probability and confidence) for sequence CCD.

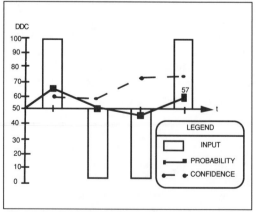

Figure 5. Situation assessment results (probability and confidence) for sequence DDC.

recency. It supports the hypothesis advanced by Einhorn and Hogarth as to the presence of a recency effect in any mixed evidence sequential information integration process.

Such effects are not predicted by classical Bayesian decision theory.

Observation 2: The confidence level in the reliability of information sources is a function of the consistency of the source. This supports hypothesis 1.

This is a short-term trend pertaining to confidence levels over time. Figure 6 collapses across the four sequences to show average confidence levels as a function of the consistency of the information source. As shown in Figure 6, commanders will upgrade their confidence levels in an incoming information source when the source provides two sequential pieces of evidence pointing in the same direction and downgrade their confidence levels when the source provides two sequential pieces of evidence pointing to opposite directions.

This finding supports our first experimental hypothesis stating that measures taken in subsequent planning cycles are interdependent. Confidence measures are a case in point. A one-cycle backtracking is necessary to be able to predict the level of confidence in an information source expressed by the commanders in the present cycle.

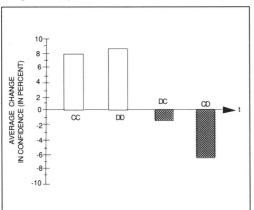

Figure 6. Change in confidence level as a function of the consistency of the source.

Observation 3: Strength of belief is only weakly related to confidence in the incoming information. We found little support for hypothesis 2, that process measures (i.e., probability or strength of beliefs) are related to the commander's level of confidence in his infor-

mation sources. Tests show the hypothesis is confirmed only for the initial observation. Recall that the intelligence messages included the intelligence analysts' own estimate of their confidence (low or high) in the information being presented. After the initial briefing in the low-confidence message condition, the commanders' average level of confidence in the information was 55.2 percent and their average probability level (strength of belief) was 61.2 percent. For the high-confidence message condition, the average confidence levels and probabilities after the initial briefing were 73.6 percent and 74.4 percent, respectively.

In subsequent stages this initial effect fades away, due to a strong recency effect and to the relative stability of the confidence levels. Surprisingly, confidence levels stabilize at about 67 percent (a mean of 66.9 percent with a standard deviation of 6.1 percent) across conditions and subjects. In other words, for a binary hypothesis-testing process, subjects assign a subjective likelihood ratio of about 2:1 that the incoming information is true. This result is consistent with previous studies of sequential information processing [16].

Planning Results

The following observations were derived from empirical analysis of the planning process:

Observation 4: Commanders tend to concentrate their planning activity in the initial planning stage, with few changes in the subsequent stages even in light of conflicting new evidence.

If we measure the total amount of unit movement and assignment at each stage regardless

Order \ Step	1	2	3
CCD	5.9	0.5	1.6
CDD	4.5	0.5	1.4
DCC	4.8	1.9	1.0
DDC	4.6	0.2	1.6
Mean	5.0	0.8	1.4

Table 2. Average number of blue units moved or reassigned at each stage of the planning process.

of the direction of the troop commitment, we can construct a table (Table 2) representing the total quantity of activity at each stage for different sequences of incoming evidence. The pattern is quite significant (p < 0.02): most of the planning activity is concentrated in the first stage.

This seems to represent a *primacy*-oriented planning behavior different from the one observed in situation assessment where *recency* was the central factor. It seems that the experienced commanders are quite willing to modify their opinions or beliefs, sometimes drastically, in the light of new evidence, but are reluctant to move from the position to which they anchored in the initial stage [12]. Their elaborate mental model of the battlefield probably includes the high cost of moving troops as opposed to the relatively low cost of changing a belief in a hypothesis. In other words, they recognize that the dynamics of action is different from the dynamics of information. This very interesting hypothesis should be further tested in realistic headquarters environments.

Observation 5: Commanders do not commit themselves to a single hypothesis but rather hedge against future uncertainties. This supports hypothesis 1.

Most (82 percent) of the commitment of units follows one pattern: a movement or reassignment of units from the neutral area (central or division command) toward the two likely avenues of enemy attack. Although most of the units were reassigned to the most likely avenue of attack, some were moved in the opposite direction in what appears to be a hedging maneuver. Figure 7 illustrates this finding.

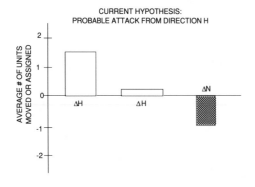

Figure 7. Representative decomposition of a commander's planning decision.

Summary of Results

Our main findings demonstrate that even in the simplest C^2 environment, situation assessment and planning are *dynamic* processes. The dependent variables of one cycle become the independent variables of the next and the derived measures are closely interrelated. Our experimental findings indicate that the order of the new incoming information has a strong effect on situation assessment. This order effect produces a recency-type behavior for the situation assessment process, with a higher sensitivity to negative evidence. For the planning process, a primacy-type behavior was observed for all evidence sequences. It was determined that the commander's confidence level in incoming information is sensitive to the consistency of that information. A surprising observation is that this confidence level tends to stabilize around the value of 67 percent (equivalent to a likelihood ratio of 2:1 for incoming evidence) regardless of the context of the information. The next section will build on these findings to develop models of the situation assessment process.

Modeling Results

We evaluated three candidate models to explain and predict the sequential belief updating process observed during situation assessment. The drastic effect of the order of incoming evidence on the direction-of-attack assessment must be accounted for in a realistic model. We considered three models: a descriptive anchoring-and-adjustment model developed by Einhorn and Hogarth, called the contrast-inertia model, a normative Bayesian model, and a normative-descriptive modified Bayesian model.

Descriptive Contrast-Inertia Model

This model, a variation on Einhorn and Hogarth's 1987 contrast-inertia model [13], is based on three principles concerning the way that information is processed. It assumes an anchoring-and-adjustment process; the evaluation of a single hypothesis (or an exclusive binary hypothesis, e.g., North-South) in the light of confirmatory (positive) or disconfirmatory (negative) evidence and the existence of a conflict between forces of adaptation and inertia.

We use the following notation:

p_i probability (degree of belief) at stage i that the attack will come from the ini tially assumed direction

p_o initial degree of belief in the initial hypothesis

a   sensitivity to disconfirmatory evidence $0 \le a \le 1$

b   sensitivity to confirmatory evidence $0 \le b \le 1$

C_i subjective confidence in the ith piece of evidence

The general form of the model is therefore:

$$p_i = \text{function } (p_{i-1}, \text{ a or b}, C_i) \qquad (1)$$

Two forms of the model distinguish the processes for dealing with disconfirmatory and confirmatory evidence. Evidence is subject to a contrast effect, that is, its impact reflects prior position. Specifically, the model supports the hypothesis that negative evidence will be discounted more when the preceding degree of belief is large, whereas positive evidence will lead to greater upward revision of belief when the degree of belief is small. The same negative evidence will induce greater discounting when prior opinion is high as opposed to low, and the same positive evidence will have more impact on belief when prior opinion is low as opposed to high. These contrast effects can be modeled by setting the adjustment weights for negative and positive evidence proportional to, respectively, the prior degree of belief and its complement. Thus,

$$p_i = p_{i-1} - a \cdot C_i \cdot p_{i-1}$$

if the ith piece of evidence is disconfirmatory (2)

and

$$p_i = p_{i-1} + b \cdot C_i \cdot (1-p_{i-1})$$

if the ith piece of evidence is confirmatory (3)

The force toward inertia can be captured in the above formulation by considering the meaning of the constants of proportionality, a and b. We define a and b to represent sensitivity toward negative and positive evidence, respectively. Small values of a and b imply low sensitivity to new information; large values of a and b imply high sensitivity. The parameters a and b model the force toward inertia by

providing a mechanism for dampening the contrast effect.

The contrast-inertia model provides an elegant mechanism for predicting the order effects evident in the experimental data. Why did information processed last in a mixed sequence of information updates (e.g., CDC) have more effect on final opinion than information encountered earlier? The answer can be easily proven in a mathematical manner by exercising Equations 2 and 3. The model always predicts recency effects in the sequential evaluation of mixed evidence, a fact that was observed in all our data-gathering trials.

In terms of overall performance, the model predicted the human data for all four sequences very well. We assumed an average C_i of 0.67, based on the confidence levels observed in the experiment, as discussed above. The best values for the sensitivity coefficients a and b were chosen to be 0.48 and 0.38 respectively. Note that the sensitivity to negative evidence, a, is larger than the sensitivity to positive evidence, b. For this set of values, the model's predictions of updated probabilities stayed within 11.2 percent of the subject's average. This is a remarkable fit if one takes into account the relatively large inter-subject variability (20.4 percent).

Normative Bayesian Model

The Bayesian framework is a normative model for sequential inferences, hypothesis updating, or revision of beliefs. It provides a means to combine new information with an existing hypothesis and produces the resulting hypothesis in an optimal fashion. The literature is full of examples showing that humans are not truly "Bayesian" and are affected by various limitations and biases that prevent them from integrating information in an optimal fashion [13,16]. Nevertheless, Bayesian models provide a frame of reference for comparing actual human decisions with optimal behavior.

The Bayesian formulation, unlike the previous contrast-inertia model, does not allow for recency factors that could produce order effects. The Bayes formula is a batch processor. At the end of a sequence of mixed evidence, only the amount and quality of information that went in is important, not the order.

How well did the Bayesian model predict the

subjects' data? For reasons mentioned before and for comparability with the contrast-inertia model, we fixed the likelihood ratio at 2:1 for confirming evidence and 1:2 for disconfirming evidence. This reflects the empirical average confidence level of 0.67 from the experiment. The Bayesian predictions were, on average, within 25 percent of the subjects' averaged data (as compared to 11.2 percent for the contrast-inertia model). The Bayesian model showed a "reluctance" to downgrade probabilities in the presence of negative evidence, while the subjects and the contrast-inertia model were over-sensitive to negative evidence.

Normative-Descriptive Model

The first step in normative-descriptive modeling, following the principles described in [17], is to develop a normative model. In this case, the normative model is a purely sequential Bayesian scheme to optimally determine the probabilities (p_i) as described above. The normative model makes these determinations based on evidence derived from the decision environment (intelligence messages, maps, etc.). Following the formal delineation of the normative scheme, qualitative descriptive factors deduced from the statistical analyses of the subjects' strength-of-belief data are introduced into the normative formulation so that it becomes more descriptive of the human subjects' data. The parameters of this normative-descriptive model are tuned until the best possible fit between the model's output and the subjects' data is achieved.

In our experiment, the most salient descriptive feature is the recency factor that produced an order effect in the sequential probabilities. This recency factor is in part due to the asymmetry in the subjects' reactions to positive and negative evidence. Subjects tended to react to negative evidence more drastically and to positive evidence less drastically than the normative Bayesian model predicts.

The Bayesian formula may be modified to reflect this phenomenon. The Bayesian theorem may be written as follows:

$$\frac{p_i}{1-p_i} = \frac{p_{i-1}}{1-p_{i-1}} \cdot LR_i \qquad (4)$$

where LR_i is the likelihood ratio of the incoming i^{th} piece of evidence:

$$LR_i = \frac{p(\inf_i / H)}{p(\inf_i / \overline{H})} \qquad (5)$$

Following Edwards et al. [9], a normative-descriptive version of Equation 4 is:

$$\frac{p_i}{1-p_i} = \frac{p_{i-1}}{1-p_{i-1}} \cdot (LR_i)^g \qquad (6)$$

where

$0 < g < 1$ for positive evidence ($LR_i > 1$)

$g > 1$ for negative evidence ($LR_i < 1$)

This manipulation reduces the effect of positive evidence while enhancing the effect of negative evidence, thus reflecting human-like behavior. The values of $g = 0.9$ for positive evidence and $g = 1.3$ for negative evidence achieved the best data-matching scores. The resulting normative-descriptive model predicted the data with an average error of only 13.1 percent, comparable with the performance of the contrast-inertia model. Note that both models capture the asymmetry of the subjects' sensitivity toward negative evidence.

Conclusions

This study examined some basic characteristics of a headquarters planning process in a focused experiment. By using a static (paper-and-pencil) task environment rather than a dynamic war gaming computer-based scenario, a small sample of experienced commanders rather than hierarchies of commanders and their subordinates, and model-based experimental methods rather than exclusively empirical evaluation, we hoped to draw conclusions and formulate hypotheses for future larger-scale experiments and exercises. Can such focused, cost-effective mini-experiments be used as "rapid prototypes" to gain insight into the HEAT cycle process? We believe that the present effort has shown the answer to this question to be positive.

Three HEAT-related hypotheses were tested in the study. The first hypothesis was confirmed. Process measures such as strength of beliefs or troop activity were found to be strongly interdependent from cycle to cycle.

Empirical evidence as well as predictive modeling showed how the very nature of C^2 decisions is dynamic. For example the strength of one's current belief that one's hypothesis is true is a function of past beliefs, new evidence and the history of the information source providing this evidence.

The second hypothesis was only partially confirmed for some instances of the situation assessment process. Although all the models tested predicted that the commanders estimate of the situation should depend on his confidence in the information sources, we were not able to show a clear dependency. More research must be done by investigating the planning activity as a function of a commander's current beliefs and confidence in understanding the battle.

The third and final hypothesis was strongly confirmed. The order in which information was presented to the commanders had a very significant effect on their situation assessment behavior. Both the contrast-inertia descriptive model and a modified Bayesian normative-descriptive model were able to predict the commander's situation assessment behavior remarkably well.

In conclusion, the first and third hypotheses provide strong support to the proposed view of the headquarters planning process as "rolling." The rolling plan concept puts the headquarters' decision processes in a dynamic perspective, where equivalent attention is being given not only to the amount and quality of the incoming evidence, but also to the sequencing of that evidence. Training commanders in a more balanced way of integrating this information may produce more effective command and control planning decisions.

This empirical support for the rolling plan concept suggest a direction for expansion and revision of the current HEAT Adaptive Control paradigm and the measures based on it. The current paradigm views the planning cycle as a succession of independent processes. This conclusion is evident in the omission of measures for correlating observations from one planning cycle to the next in the current set of HEAT measures. The existing HEAT methodology cannot reveal how decision makers learn about their environment, select options to facilitate learning (e.g., probing), use learning in a feedback fashion to select options, or hedge

decisions in anticipation of courses of action they may pursue in the future. Further HEAT development should concentrate on measures that capture the interdependency of successive planning cycles.

References

[1] Lawson, J.S., "Command and Control as a Process," *IEEE Control Systems Magazine*, March 1981, pp. 5-12.

[2] Lawson, J.S., "The Role of Time in a Command and Control System," *Proceedings of the Fourth MIT/ONR Workshop on Distributed Information and Decision Systems Motivated by Command-Control-Communications Problems: Volume IV*, Cambridge, Massachusetts: Laboratory for Information and Decision Systems, LIDS-R-1159, MIT, pp. 20-59.

[3] Wohl, J.G., "Force Management Decision Requirements for Air Force Tactical Command and Control," *IEEE Transactions on Systems, Man, and Cybernetics*, Vol. SMC-11, No. 9, September 1981, pp. 618-639.

[4] Wohl, J.G., E.E. Entin, D.L. Kleinman, and K. Pattipati, "Human Decision Processes in Military Command and Control," in *Advances in Man-Machine Systems Research*, Volume 1 (A. Sage, ed.), JAI Press, 1984, pp. 261-307.

[5] Defense Systems, Inc., *Theatre Headquarters Effectiveness: Its Measurement and Relationship to Size, Structure, Functions, and Linkages*, Volume I, McLean, Virginia, December, 1982.

[6] Cushman, John H., Personal Communication to Robert R. Tenney, 1988.

[7] Wohl, J.G., "Rate of Change of Uncertainty as an Indicator of Command and Control Systems Effectiveness," *Proceedings of the Fourth MIT/ONR Workshop on Distributed Information and Decision Systems Motivated by Command-Control-Communications Problems: Volume IV*, Cambridge, Massachusetts: Laboratory for Information and Decision Systems, LIDS-R-1159, MIT, pp. 1-17.

[8] Bertsekas, D.P., *Dynamic Programming: Deterministic and Stochastic Models*, Englewood Cliffs, New Jersey: Prentice-Hall, Inc., 1987.

[9] Edwards, W., and L.D. Phillips, "Man as Transducer for Probabilities in Bayesian Command and Control Systems," in M. W. Shelly, II, and G. L. Bryan (eds.), *Human Judgments and Optimality*, New York: John Wiley, 1964, pp. 360-401.

[10] Peterson, C. R., and L.R. Beach, "Man as an Intuitive Statistician," *Psychological Bulletin*, 69, 1967, pp. 29-46.

[11] Donnel, M. L., and W. M. Ducharme, "The Effect of Bayesian Feedback on Learning in an Odds Estimation Tasks," *Organizational Behavior and Human Performance*, 14, 1975, pp. 305-313.

[12] Tolcott, M.A., F. F. Marvin, and P. E. Lehner, "Effects of Early Decisions on Later Judgments in an Evolving Situation," Technical Report No. 87-10, Falls Church, Virginia: Decision Science Consortium, Inc., 1987.

[13] Einhorn, H.J., and R.M. Hogarth, "Behavioral Decision Theory: Processes of Judgment and Choice." *Annual Review of Psychology*, 1981, 32, 53-88.

[14] Einhorn, H.J., and R.M. Hogarth, "Adaptation and Inertia in Belief Updating: The Contrast-Inertia Model," Technical Report, Center for Decision Research, Chicago: University of Chicago, 1987.

[15] Entin, E.E., R.M. James, D. Serfaty, and J. Forester,

"The Effects of Cognitive Style and Prior Information on Multi-Stage Decisionmaking," Technical Report Number TR-277-1, Burlington, Massachusetts: ALPHATECH, Inc., 1987.

[16] Lopes, L.L., *Averaging Rules and Adjustment Processes: The Role of Averaging in Inference (T.R.)*, Madison, Wisconsin: Wisconsin Human Information Processing Program (WHIPP 13), 1981.

[17] Bushnell, L., D. Serfaty, and D.L. Kleinman, "Team Information Processing: A Normative-Descriptive Approach," in *Science of Command and Control: Coping with Uncertainty*, Johnson and Levis, (eds.), Fairfax, Virginia: AFCEA International Press, 1988.

10

Planning for Integrated System Evaluation: An Application to SDI

Michael F. O'Connor

This paper discusses planning for integrated system evaluation and is based on both theoretical and empirical evaluation of Battle Management/Command, Control, and Communications (BM/C³) architectures for the United States Army Strategic Defense Command (USASDC) and the Strategic Defense Initiative (SDI) program. The concepts are relevant to any program evaluation involving projecting performance of a major system. The work is prescriptive and methodological in nature, discussing what should be done rather than featuring, for example, software designed to solve part of the problem. Nonetheless, implementation of the ideas discussed here does not involve the development of major new algorithms or software. Much of what is discussed has been termed "good engineering design" by one researcher, who indicated that these ideas are routinely accomplished. However, this approach expands the meaning of "good design." What the researcher implied was being routinely implemented, is not, perhaps for reasons of increased cost, schedule slippage, programmatic constraints, the usual organizational approach to system acquisition, and, in some cases, laxity.

The approach here called a top-down approach to integrated system evaluation, involves linking user, procurer, and designer in a trade-off analysis that is broader in scope. The analysis is to be done early in the program, using whatever tools are available. Difficulties in accomplishing this approach are organizational, not technical, and though parts deal with the use of complex performance simulations, the perspective adopted for system acquisition decisions is one of an organizational decision maker. Put theoretically, the utility function relevant to the design decisions is a social one, involving many organizations. Put practically, the information required to accomplish design trade-off analysis resides within several organizations and requires a systematic integration of trade-off analysis guided by an overall evaluation plan. Such trade-off analyses necessarily involve more than subsystem performance trade-offs addressed in complex simulations.

This presentation will involve three phases: the theory of the approach, the methodology involved, and an example of procedures to be followed in implementation.

A Theoretical Framework for Trade-off Analysis

Any decision with clearly defined options involves uncertain future events and value trade-offs with respect to consequences. The system design problem is further complicated because the decisions do not involve fixed options, but rather the design, development, testing, and deployment of systems that must meet performance goals in uncertain future scenarios. They also must satisfy numerous societal criteria such as low life-cycle cost and political acceptability.

The traditional representation of a decision problem involves a decision tree in which relevant outcomes are specified as branches terminating in consequences uniquely associated with the particular branch. Figure 1 shows an alternative representation of the decision problem that is particularly suitable for discussing the system acquisition problem. Four abstract spaces are represented. (This approach is based

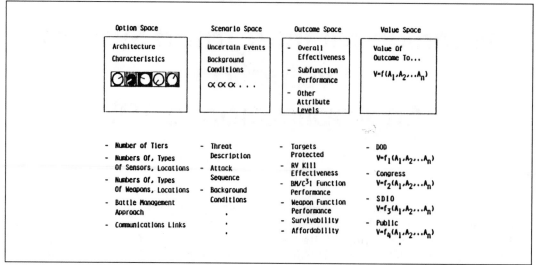

Figure 1. Four abstract spaces.

on work presented in O'Connor and Edwards, 1977 and O'Connor, 1985). Option Space contains all possible designs of the kind in question. For the SDI problem, it is the space of all architectures and associated characteristics. It can be thought of as a vector space; but however viewed, it contains all the characteristics relevant to specification of an architecture. This is the space of the engineer or designer who specifies an architecture by selecting a set of levels of characteristics. The specification of an architecture is displayed as settings on several dials. In Option Space, it is acknowledged that design dials cannot be independently manipulated, because a change in one can alter others. Designs can be very specific or general high-level designs, and a set of specific characteristics can be mapped into more general system characteristics. Thus, design options can be represented in hierarchies. The main point is that all design decisions and changes occur as changes in Option Space.

Once a system architecture, such as a BM/C³ architecture is specified, it can be evaluated by considering all futures in which the system will operate. These futures are characterized in Figure 1 as scenarios in Scenario Space. As with Option Space, this also can be viewed in several ways: as a vector space; as an event tree whose branches comprise the set of all future scenarios; as a sample space; or as some combination thereof. A scenario is viewed as consisting of two parts: the stage setting into which the system is introduced and the subse-

quent unfolding of acts and events that detail the future. Many scenarios are possible and all potential futures for a system would comprise an infinite set. Care must be taken to develop an efficient, manageable set of scenarios that will allow for valid system evaluation and that also will be cost-efficient. Such a set of scenarios should represent the future in which the system will be deployed, while at the same time discriminating among alternative system designs in terms of value differences. These two requirements, coupled with the need for efficiency, will most likely be in conflict: the first yielding broad, and the second fairly specific, scenarios.

The value differences are represented as levels of attributes in Outcome Space. Outcome Space also can be visualized as a vector space. Attributes in Outcome Space can be types of performance; different "ilities" such as survivability; or different attributes such as "world political stability." Outcome Space contains all attributes relevant to specification of the outcome of an Option Space/Scenario Space combination. One can think of these attributes as structured in a hierarchy in which descending levels contain sub-attributes of increasing specificity.

The Value Space illustrates the important fact that the same outcomes have different values to different populations. For the SDI example, diverse interest groups such as DOD, Congress, the general public, and the Soviet Union have different values for the same set of

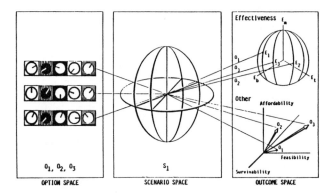

Figure 2. Mapping: Option space to outcome space.

outcome levels. Value trade-offs among the attributes of Outcome Space will be different for these populations. Thus, the same point in Outcome Space can map into a large number of points in Value Space.

Another important point with respect to system design, often neglected by those who emphasize performance simulation as the major aspect of system evaluation, is that the same levels of overall system performance can be achieved by different systems having different values for other attributes such as affordability, feasibility, and survivability (See Figure 2). While this is no revelation to system evaluators, what to do about this has usually been overlooked. These attribute trade-offs are not after-the-fact evaluation issues. Rather, they are decision issues relevant to the design problem and should be resolved before detailed system design and testing take place. This is difficult and involves good mission analysis (often neglected) and policy analysis which addresses those complex value trade-offs characterized by Outcome Space/Value Space combinations. Which is the best way to achieve a given desired level of performance? One way would be to deploy a large number of lower performance platforms representing a near-term solution. A second way would be to opt for a more risky, less feasible but higher performance technical solution, with fewer required platforms that still achieve the same performance levels.

What about political aspects? How much of this should be examined up front? For SDI, this issue lies with the policy analysts whose domain involves developing a system where the options are driven by political differences. One thing seems clear: Value trade-off ques-

tions must be resolved early in the design process. A decision must be made with respect to those attributes relevant to evaluation of this design problem. The development of measurement scales for these attributes must also be addressed.

The message of this introduction is simple: value trade-offs relevant to system development should be addressed early in the design process and in a methodical fashion that yields design direction. Evaluation tools are required for this to be done. As with most other major system developments, SDI performance simulations will be necessary. Many simulation models exist. Each answers some part of the performance simulation problem, but none does the complete job. Generally, the more complicated the simulation, the more expensive it is to run, change, and understand. But the system evaluator need not shrink from these considerations. Rather, he should plan to do what he will eventually be forced to do: that is to use what is available when he has to make his decision. How to do this will be discussed in the following sections.

The Methodology

Evolving Architectures

Table 1 illustrates the approach to evolve architectures. The process involves specifying a relevant scenario (Step 1); specifying the required effectiveness scenario (Step 2); and identifying an initial architecture (Step 3), which will yield the desired effectiveness. That architecture is then "deployed" in the scenario and the remaining details (part of Step 4) of the scenario are developed. This can be done using a computerized simulation to esti-

IN OPTION SPACE	IN SCENARIO SPACE	IN OUTCOME SPACE	IN VALUE SPACE
	1. Fix Scenario to S1	2. Determine Required Effectiveness in S1	
3. Identify an Initial Architecture	4. Deploy in Scenario S1 and Generate Resultant Scenario Detail.	5. Assess Functional Performances; Verify Overall Effectiveness. Assess Other Attribute Levels.	6. Assess Value. Prioritize Potential Improvements
7. Assess feasibility of improvements.	8. Repeat Step 4 with new architecture	9. Repeat Step 5 with new architecture	10. Repeat Step 6 with new architectures
11. Repeat Steps 7-10 until no further improvements are needed or desired.			
	12. Fix Scenario S2	13. Determine Required Effectiveness in S2	
14. Repeat Steps 7-11 for Scenario S2			
15. Repeat Steps 12-14 for Scenarios S3 through Si			
16. Finalize Preferred Architecture(s). Record details of all trade-off analyses, scenarios, and architectures.			

Table 1. Using the approach in architecture design.

mate performance, or by using a judgmental approach such as step-through simulation (See Ulvila, J. and Brown, R. V., 1978). The assessment of system performance and other attribute levels is accomplished in Outcome Space. Part of this involves verification that the proposed system actually yields the desired performance. The detail to which performance can be modeled at this point is an issue of paramount importance. The goal is not to obtain two-digit or better performance estimate accuracies, but rather to examine value trade-offs in detail. This will involve intersystem performance trade-offs as well as trade-offs between performance and other attributes such as cost, political stability, etc. Initial design studies will deal with high-level attribute differences, and performance need only be systematically related to the precision levels of these differences.

By examining attribute levels associated with the proposed system including cost, "ilities," political, social, and other outcomes, potential improvements are identified and prioritized (Steps 5 and 6). The architecture is then reworked with the designers and improvements (with respect to attributes) are sought (Steps 7 through 10). Note that this can be an exploratory policy analysis at this point. Analyses can be judgmental. Performance estimates can be expert assessments, because architecture differences that yield different attribute levels with respect to major political and social issues will be major differences. Detailed performance assessments may not be required, even if sufficient detail exists to make

such estimates.

Alternatively, such analyses may require detailed performance simulation. For an SDI example, consider an architecture that yields acceptable leakage protection in a scenario, but is designed so that its surveillance system and weapons are deployed at such low levels as to be vulnerable to antisatellite weapons. Suppose that the performance simulation does not address satellite attack. Then several different architectures can be designed, each yielding different leakage performance, and having different vulnerabilities. The two attributes are assessed separately and then addressed in value trade-offs. Which is better; increased performance with increased vulnerability or vice versa? Depending on simulation complexity, such off-line analyses will be required.

It is easy to see that this early trade-off analysis, involving an approach such as Multiattribute Utility Assessment (MAUA—see O'Connor, 1985), can be of different levels of generality and can also be an exploratory, iterative analysis involving "excursions in Outcome Space." For example, the implications can be explored for architecture design stressing early technical feasibility, survivability, political acceptability, low lifecycle cost, graceful degradation, etc. The crucial point is the link between Outcome Space/Value Space trade-off analysis and design implications.

The level of precision need not be that of a simulation that estimates, for example, leakage in terms of the number of reentry vehicles (RVs). Rather, meaningful measurement scales for attributes should be used, and they can be defined as relative to the attribute levels of the architectures under consideration. As more precision and detail are required, more specific measurement devices may be required. This will be discussed later.

Once the architecture trade-offs have been examined for this first scenario, the same process must be repeated for the other relevant scenarios (Step 12-15). This is most relevant to performance estimates, although changes in scenario can also affect levels of other attributes. An example is survivability, if it is not part of the performance assessment as discussed earlier. Another example would be political acceptability.

Note that any architecture changes that would yield attribute levels in subsequent scenarios will potentially affect attribute levels in earlier scenarios, and checks must be made to ensure that this has not occurred. If it has, it must be noted, and either a compromise design must be achieved, or a decision must be made to go with one or the other of the designs in question. Thus, interscenario performance trade-offs are yet another aspect of the organized trade-off analysis.

Most of the iteration between scenarios and the impacts of design changes across scenarios will involve details relevant to performance assessments, fairly specific parameter differences such as threat types, launch sequence, warning time, etc. Other attribute trade-offs, relevant to such issues as policy analysis, will often require fairly broad scenario differences. For example, different scenario details will be relevant to discriminations among architectures with respect to issues such as political acceptability and treaty compliance. However, the total evaluation should remain connected, and assumptions on scenarios and constraints introduced should be consistent across analyses. (For a detailed discussion see O'Connor and Edwards, 1977.)

The Use of an Attribute Bank

The issue of quantification of attributes in trade-off analysis, measures of effectiveness, and the validity and reliability of subjective value assessments are familiar to practitioners of decision analyses. Decision analyses involving one-time evaluation of options usually quantify attributes using relative intra-attribute value scales having ranges determined by the specific options under consideration. Those who seek global measures of "effectiveness," value, or utility, to be applied more than once, must find intra-attribute value scales with ranges that include all potential options. This can be a difficult problem and, thus arises the issue of relative versus absolute measurement scales, or in measurement theory, the related issue of the uniqueness problem. It is fairly simple to determine an appropriate measurement scale for the height of basketball team candidates. It is quite another to define a measurement scale for the attribute known as BM/C^3 system effectiveness.

The focus of this paper is not the establishment of measures of effectiveness for BM/C^3 systems, although that was a required part of the actual evaluation. The focus is on the use of trade-off analysis in the design process, and that necessarily involves quantification of attributes. As noted earlier, the evaluator must use the tools at his disposal. Often too much concern is placed on the so called "apples and oranges" problem, and large amounts of time and money are spent developing detailed performance simulations with all the properties necessary to address the relevant trade-off issues. In the application used as an example for this paper, numerous simulations of parts of SDI system performance were available, but none directly represented the specific BM/C^3 to system (weapons and sensors) trade-off issues in the BM/C^2 architecture evaluation. Because time was not available for simulation development, there was no choice but to proceed using the tools available at the time. The attribute bank was developed as a means of organizing that evaluation, and, at the same time, attacking the so- called "apples and oranges" problem.

The attribute bank consists of a hierarchy of potential evaluation attributes, from aggregated attributes at the top, to successively more specific measures at each hierarchy level. Similarly, the hierarchy contains simulated measures to the left and expert assessments to the right. Thus, the uppermost left corner will have aggregated, simulated measures and the lower right has expert judgments on highly specific, disaggregated attributes. Although this discussion is most amenable to performance assessments, it is also relevant to the broader trade-off analyses required for the design process.

Figures 3 and 4 illustrate some of the hierarchy for evaluating the worth of BM/C^3 architectures. Figure 3 represents the highest level whereas Figure 4 decomposes the node BM/C^2 system effectiveness. For each node in the hierarchy, a definition was developed with an accompanying discussion of measurement details in an evaluation handbook (See O'Connor, 1986).

Performance simulations were available for many of the attributes, and the outputs were characterized alternatively as measures of system effectiveness (MOSE), measures of effectiveness (MOE), measures of performance (MOP), or simply expert assessments. Examples of simulation outputs are shown in Table 2. Analysis or expert assessments may be required for other attributes. The hierarchy

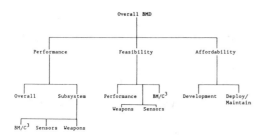

Figure 3. Overall BMD: Decomposition.

Figure 4. BM/C^2 system effectiveness decomposition.

integrates all aspects of the effectiveness evaluation and focuses trade-off analyses.

In evaluating performance, an overall measure is desirable such as that for overall BM/C^3 performance. However, any simulation capable of assessing such a complicated, aggregated performance will be detailed and difficult to understand. The evaluator must understand the simulation or have a translator, who can convert model outputs to language the evaluator can follow. If the evaluation is driving the simulation and not vice versa (which is often sadly the case) the outputs will answer evaluation questions. Trade-off analysis, based on the attribute bank structure, is to assure that this happens.

Any simulation, however good, will omit certain aspects of system worth, but the attribute bank will contain nodes indicating that aspect of worth and a potential measure. Familiarity with the particular evaluation tools will allow mapping to the nodes of the bank that are covered and this will pinpoint attributes not covered. They can be separately assessed through analysis or judgment and the results can be combined with simulation outputs to yield a complete evaluation. Often decision-analytic trade-off analysis will be required to accomplish the integration of such measures, but this is an implementation procedural issue and not a theoretical problem.

An example of this problem, is when a simulation is used to estimate system performance using leakage. However, survivability aspects of certain components are not included in the simulation. The simulation can be run for several different architectures and the survivability can be assessed separately. Two attributes can then be defined: leakage performance without attack by the enemy, and survivability. The trade-offs between the two attributes can be assessed and architectures located on each. Note that the issues of independence in Option Space and value independence in Value Space must be carefully addressed.

Such a thorough, organized evaluation is not easy, even when using trade-off analysis as recommended here. Information like that appearing in Table 3 on evaluation issues covered, must be organized in the attribute bank so that the evaluators know what is covered, to what degree, and by what instrument. Then the evaluators have control over the evaluation

Analytical/Simulation Model Types	Modeling Examples	Output Measure Type	Output Measure Examples
Strategic Defense System (SDS)	•End-to-end system representation	MOSE	•Value of defended targets surviving • System leakage
Specific Grouping of of system elements*	• SDS midcourse tier • Specific RBM & Subordinate LBMs, weapons & sensors	MOE	•Tier-leakage •Threat capacity (for specified grouping)
System element	• BM component of AOS	MOE	•BM computer processing delays (system element level) •Discrimination capacity
Functional/Small Functional Grouping	• Discrimination • Comm. link propagation	MOP	•Discrimination data processing delays •Critical message delays
Phenomenology	• NWE • Conventional Fragmentation warhead effects	Inputs to higher level models	•Atmospheric perturbation.

• e.g., elements affecting a specific tier, region, or other defense volume.

Table 2. Performance modeling.

Issue	Level of Coverage	Models
BM/C³ processor/comm trade-offs • processor speed and location • communications throughput	• Time delays explicitly accounted for • Representation of SDS architecture at system element level • Performance-related comm. characteristics for comm. link types	TDS, SIMSTAR
Man-in-the-loop issues	• Impacts of time delays associated with manned functions	TDS
Survivability of SDS elements	• Temporary or permanent damage due to hostile activities	TDS
Subsystem performance capability trade-offs • sensor discrimination • IFGU requirements	• Impacts of specific system element capabilities on engagement timelines, battlespace	TDS, DIOS, COVER, SAS library

Table 3. Evaluation issue areas.

issues and can use the expertise available to them in a continuation of the organized trade-off analysis.

The evaluator uses the attribute bank to pick the relevant issues. The mapping of performance models to the attribute bank permits a choice of attributes to be assessed, using performance models, analyses, or expert assessments. Where a model lacks coverage of a specific aspect of system design and related performance, assessments are obtained and combined either judgmentally, using a form of trade-off analysis, or by an integrated application of available models. The latter process is further described below.

Procedures Used in Applying the Methodology to the SDI BM/C³ Evaluation Problem: An Example _____

The Procedures

In this section, we will address the difficult problem of assessing the worth of different BM/C³ structures and how the trade-off analysis should proceed, using the tools that are available. In Figure 5, the BM/C³ to systems iterative integration process is illustrated. A problem with SDI BM/C³ evaluation, and with C³ evaluation in general, relates to the "force multiplier" problem. "C³ does not kill people, bullets do!" This argument is old and need not be discussed here save to indicate that most simulations emphasize weapon and sensor characteristics and configurations while minimizing representation of BM/C³ characteristics, partly because of the difficulty of including all of these details in one simulation. The system dynamics representation for a problem like SDI

is already huge for simply the weapon and sensor performance modeling without considering the many complicated details of BM/C³.

Thus, it is usually the case in BM/C³ design, that weapon and sensor parameters are constraints on the BM/C³ system and not vice versa. Figure 5 represents an alternative approach in which reasonable bounds on the BM/C³ system serve as constraints which limit the range of allowable weapon and sensor parameters, and, thus, the BM/C³ becomes the "driver" of the design optimization. Given the great uncertainty with respect to BM/C³ feasibility, this is, indeed, appropriate.

The guidelines developed for BM/C³ by the well-known Eastport Study Group are listed below. This group was particularly concerned with software and algorithm feasibility.

• BM/C³ must be integral to overall architecture, not an applique;
• Principal influence on BM/C³ design must be reducing software complexity;
• Most plausible BM/C³ architecture is hierarchic;
  - decentralized C²
  - battle groups with local autonomy
• BM/C³ system should be open and distributed;
  - loose coordination
  - maintain central command authority/global situation assessment
• BM/C³ components should be simple and redundant, even if suboptimal;
• BM/C³ testing should be realistic;
  - infer full-scale performance through test of small segment
  - use simulation extensively

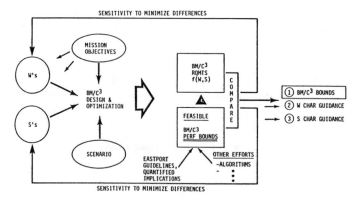

Figure 5. BM/C³ to systems iterative integration process.

- develop an in-line testing
  capability

• BM/C³ architects should consider establishing a separate, dedicated communications network.

As indicated, that group argued that BM/C³ should be an integral system component of the design and not simply an applique added after thorough consideration of weapons and sensors. The evaluation problem then becomes one of ascertaining the impact of these recommendations, if adopted, on performance and associated systems. While it was possible to develop some off-line assessment of BM/C³ performance impact and integrate this with other measures using trade-off analysis as discussed earlier, it was believed that a more integrated assessment of weapon-sensor/BM/C³ performance trade-offs was required. The question to be answered is displayed in Figure 6. Given definition of the axis labeled as BM/C³ simplicity, which was the correct curve? The dotted line represents required performance in the scenario, and it is necessary to decide where on the BM/C³ continuum to locate to assure that requirements are met. While numerous simulations of total system performance were available for the evaluation, none of those available provided detailed representation of overall SDI system performance (e.g., RV. leakage) as a function of weapon details, sensor details, and similar levels of detail with respect to BM/C³ characteristics, such as organizational structure, communications throughputs, BM/C³ processor throughputs, etc. Those that were large, detailed, multi-tiered models generally provided detailed dynamic representation of threats, weapons,

and sensors, but little or no representation of BM/C³ characteristics. (For example, the available version of Defense in Depth Simulation [DIDSIM], assumed instantaneous communications between nodes and a centralized BM/C³ management structure. Later versions were under development to change these simulation characteristics.) How could the evaluation be accomplished with the tools at hand?

One procedure for the evaluation problem described involves assuming that a simulation of overall performance is available that accurately portrays weapon and sensor characteristics and resultant performance impacts. (Numerous detailed simulations, e.g., DIDSIM, were available at the time.) The first step in the procedure is to optimize the system in terms of numbers, locations, and capabilities of weapon and sensor parameters in detailed performance trade-off studies. Impacts of major weapon and sensor characteristics are identified and quantified (e.g., midcourse tier leakage). At this point, BM/C³ characteristics have not been introduced. BM/C³ essentially is assumed to be a "perfect" system, that is, no BM/C³-induced performance decrements are assumed.

Next, the attribute bank is used to develop a set of BM/C³ related attributes to be varied in the evaluation. Major structure and capability impacts and a method to characterize performance impacts are identified, as well a method to characterize performance impacts. Examples of such BM/C³ attributes include centralized versus decentralized decision making, preprogrammed decision templates versus dynamic response, and proactive versus reac-

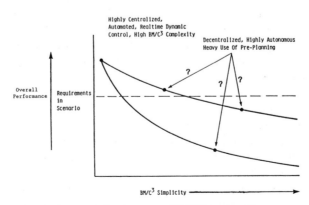

Figure 6. Performance/BM/C³ trade-offs.

tive control mechanisms. Using these details, the BM/C³ structure is developed that yields appropriate attribute levels. This is overlaid on the simulation by indicating which functions will be performed where, which information will be available to which platforms, and what threats can be engaged as a result. An example with respect to such attributes as performance, survivability, and feasibility is a highly centralized BM/C³ architecture versus a completely autonomous one and the BM/C³ management algorithms required for each. For the centralized structure, the system can utilize weapons and sensors to engage threats in an optimal overall fashion. For the autonomous architecture, each weapon must engage the threats in the order it encounters them until it runs out of "bullets." BM/C³ complexity varies dramatically for these two architectures.

At this point, only the BM/C³ management structure is modeled in terms of information available and decisions made. No delays due to communications or processors are included. Given the analysis thus far, the impact of increased simplicity should be decreased SDI system performance.

At this point, costs due to centralization are not represented. Thus far, the BM/C³ network has been assumed to cause no delays, and weapons and sensors receive information as required to operate within their respective cycles. However, the more centralized architecture will involve detailed communications and processing delays that will not be incurred in the autonomous battle groups. Thus, an estimate of such delays must be included. Another issue here is key node vulnerability and the potential for catastrophic failure of the system through system fault or enemy destruction. One of the Eastport recommendations was for a separate, dedicated communications network to avoid such a disaster. A separate communications simulation can be run to include battle managers and communications nodes and it can be used to estimate communications and processing delays due to the BM/C³ network. Although this is not the same as a complete representation of the entire network, it will suffice given the lack of the "complete" simulation. Given such a modeling effort, communications and BM/C² processor delays can be estimated and inserted into the simulations as delays in sensor and weapon performance. The

exact method for doing this depends, of course, on the simulation. The resultant system performance is then estimated. Alternatively, the required BM/C² and communications capabilities required to yield no decrement in performance can be estimated using a different analysis.

The result of this evaluation is that we can estimate the impact of sensor and weapons requirements on BM/C² and the impact of BM/C³ performance on the total architecture.

Summary

The approach described here allows the estimation of required BM/C³ characteristics given weapon and sensor parameters and it also allows estimates of system performance given "reasonable" BM/C³ parameter values derived from models of the BM/C³ network. In the implementation discussed here, it is recommended that the attribute bank serve as a guide to the procedures adopted to ensure complete and systematic coverage of all BM/C³ characteristics.

The approach described here provided for evaluation of the impact of BM/C³ architectures on the same measure of performance as that used in modeling weapon and sensor system performance. This provides a capability for modeling BM/C³ versus sensor or weapon trade-offs and also allows for completion of the evaluation within the time, funding, and organizational constraints existing at the time.

In a particular evaluation, it may be quite feasible to develop completely new simulations of performance that include all system detail relevant to architecture performance trade-off considerations. That would not change the desirability of using the attribute bank approach to integrate the evaluation and to focus an a priori trade-off analysis to identify the architecture characteristics for which key performance or effectiveness simulations are to be developed. This structured approach will give the program manager—who in the end must make the crucial decisions—control of the evaluation from beginning to end, and he will not have to rely on tools or reports that do not directly focus on the evaluation issues.

References ————————————————

O'Connor, M. F. and Edwards, W. "On Using Scenarios in the Evaluation of Complex Alternatives (Technical Report DDI/DT/TR-76-17)." McLean, VA: Decisions and Designs, Inc., 1976.

O'Connor, M. F. "Establishment of Standards and Procedures for Evaluating and Comparing SDI BM/C3I Architectures (Working Paper WP-85W00070)." McLean, VA: The MITRE Corporation, April 1985.

O'Connor, M. F. *Evaluation Handbook for the USASDC Composite Architecture Synthesis Team (CAST).* McLean, VA: The MITRE Corporation, July 1986.

Ulvila, J. W. and Brown, R. V. "Step-through Simulation." *Omega: The International Journal of Management Science*, 6(1), 25-31, 1978.-38

11

A Function-Based Definition of (C²) Measures of Effectiveness

Paul E. Girard

Introduction

Within the Space and Naval Warfare Systems Command, the Warfare Systems Architecture and Engineering (WSA&E) Directorate (SPAWAR-30) directs the development of architectural descriptions and assessments of current and future Naval Warfare Systems under the sponsorship of the Deputy Chief of Naval Operations for Naval Warfare (OP-07). In collaboration with the Anti-submarine Warfare Directorate (SPAWAR PD-80), the ASW Architecture Division (SPAWAR-315) has solicited the Naval Ocean Systems Center to lead a team of Navy laboratories to address ASW architecture. The process is initiated by the issuance of a Top Level Warfare Requirement (TLWR) by OP-07. In response, the architecture team is attempting to devise a means of providing a traceable account of the relationship between system performance and the TLWR. This has given rise to the development of an on-going methodology for Architectural description, modeling and assessment. A byproduct of this methodology is a function-based definition of measures of effectiveness. This paper focuses on how to utilize that definition to establish C² measures of effectiveness.

After a review of some motivating concepts, a summary of the methodology will be presented. The process results in a general definition of MOEs and MOPs. This will be followed by a discussion of the role of command decision making in the execution of functions. The definition of a canonic C² MOE is supported by an example and this provides motivation for application of the definition to future analyses.

Background

There appears to be a consensus among the C³ community that the justification for C³ systems must be based on the combat or mission outcome. In other words, the effectiveness of decision making (and decision support systems) has no meaning outside the context of a mission or purpose. Conversely, the historical approaches to modeling and assessment of operational systems have implicitly assumed "perfect C³." This results in optimistic forecasts of performance; not an accurate model. In other words, the expected outcome of missions cannot be properly modeled unless the effect of decision making is included. These two complementary ideas suggest a synergistic relationship between mission analysis and decision analysis.

In fact, that relationship is one of cause and effect; the mission is not executed unless it is initiated by a decision to carry it out. In other words, it is the role of command (decision making) to initiate required mission functions. This involves recognizing which functions are required or will be effective (or appropriate or authorized) and allocating available resources under his control to carry them out. Of course, the initiation must be accomplished in a timely manner, that is, early enough for the mission to be carried out before the enemy accomplishes his objectives, but not so early that more effective alternatives might be preempted. Figure 1 highlights the three motivating concepts just described, which will be incorporated in the approach.

Another concept which is depended upon heavily is one of a hierarchy of objectives

MOTIVATING CONCEPTS

> EFFECTIVENESS OF DECISION MAKING HAS NO MEANING
> OUTSIDE THE CONTEXT OF A MISSION OR PURPOSE

> EXPECTED OUTCOME OF MISSIONS CAN NOT BE MODELLED
> UNTIL THE EFFECT OF DECISION MAKING IS MODELLED

> THE ROLE OF COMMAND (DECISION MAKING) IS TO
> RECOGNIZE AND INITIATE REQUIRED MISSION FUNCTIONS
> AND ALLOCATE RESOURCES TO THEM
> IN A TIMELY MANNER

Figure 1.

(Rahmatian). Although it is not an original idea, Rahmatian provides a simple picture (Figure 2) of the interlocking role of the objectives.[1] From an arbitrary level, "what" is done is accomplished for a higher level purpose ("why"), and "how" it is done becomes a lower level "what" whose purpose ("why") is at the original level. The right side of Figure 2 recasts the hierarchy of objectives in terms of missions, functions and tasks. For a particular force or system, its functions are what it does in order to accomplish its mission. Its tasks are its subfunctions, which are performed by its parts or subsystems. Identifying the subsystems and subfunctions at this level with the systems and functions at the next lower level establishes a nested hierarchy of functions and systems associated with the hierarchy of objectives. The missions of the lower level are identical with (a subset of) the functions at the upper level and the functions at the lower level are (a subset of) the tasks at the upper level.

The Navy states its mission, functions and tasks in Naval Warfare Publication (NWP)-1, Strategic Concepts of the U.S. Navy. By law, the mission of the Navy is to "be prepared to conduct prompt and sustained operations at sea in support of U.S. national interests." The functions of the Navy are to perform power projection and sea control. Recently, the Navy has been assigned the job of sealift and an unstated objective has always been to defend the United States. NWP-1 goes on to state that the tasks of the Navy consist of warfare tasks and support tasks. Among the warfare tasks is antisubmarine warfare (ASW). Based on the hierarchy of objectives, therefore, the mission of ASW forces is to conduct ASW operations in support of power projection, sea control, sealift and defense of the U.S. What, then, are the functions of ASW Forces? This is the purview of the ASW TLWR and is the basis of the functional analysis embodied in the methodology.

Methodology

The approach (Figure 3) consists of performing a functional decomposition in each of three context-setting domains (Mission, Organization, and Resource). This decomposition is driven by the Mission Success Criteria (MSC) and Required Capabilities (RC) established by the

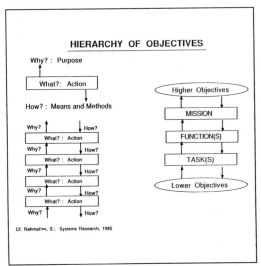

Figure 2.

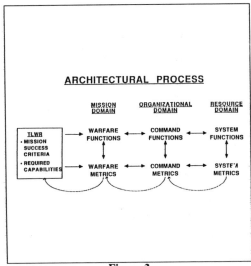

Figure 3.

TLWR. As we shall see later, the MSCs and RCs actually establish the initial tiers of functions. The functions within these domains (warfare or support mission functions, command/decision functions, and equipment or personnel functions, respectively) are mutually supportive in achieving the goals of the mission and establish a hierarchy of objectives within each domain as well as across them. Realizing that the achievement of the objective is the criterion of success establishes a one-to-one correspondence between functions and metrics. In fact, this correspondence was recognized while reviewing the MSCs and RCs. In turn, the relationship among functions also corresponds to the mathematical relationship (equations) among metrics at the same level. This concept, too, has its roots in the relationship between the MSCs and RCs of the ASW TLWR, as we shall see. This suggests the potential for finding a relationship for aggregating metrics for resources, organizations, and missions to the achievement of the TLWR objectives.

Figure 4 summarizes the elements of the ASW TLWR. For twenty-two stressing cases, the mission was stated in terms of the type of ASW mission, area (interdictive) or local (protecting other forces), and the nature (peace, crisis or war), region and timeframe of the conflict. Thus a typical mission context might be to "conduct Area ASW in the Norwegian Sea during the first phase of global conventional war." Then the mission success criteria for that

ASW FUNCTIONS

ASW TLWR

- 22 STRESSING CASES
    - AREA ASW/LOCAL ASW FORCE
    - REGION
    - TYPE OF CONFLICT
    - TIMEFRAME ⎤ MISSION

- MISSION SUCCESS CRITERIA (MSC)
    - EFFECTIVENESS GOALS
    - PROFESSIONAL JUDGMENT
    - FIRM ⎤ FUNCTION

- REQUIRED CAPABILITIES (RC)
    - PERFORMANCE MEASURES
    - PERFORMANCE COUPLING TO MSC
    - C3 NOT COUPLED
    - NEGOTIABLE ⎤ TASKS

Figure 4.

mission are stated, such as, attrite a percentage of the expected Order of Battle. These are the firm "top" level warfare requirements. The Required Capabilities were devised to exhibit a degree of credibility to the potential of achieving the MSCs. The RCs are selected factors in an equation relating several performance factors to the MSCs. Examples of RCs are kill probabilities given detection, classification and localization or the area that can be searched to achieve a specified detection probability. The equation that aggregates the RC metrics to the MSC metrics is called the audit trail.

Examining the MSCs and RCs and other elements of the audit trails reveals most can be stated in terms of the probability of successfully accomplishing a function. This suggests that a function defines its own metric, which is the probability of a successful outcome of performing the function. Conversely, a probability metric defines the outcome of its related function. For example, the probability of detection identifies the detection event as the outcome of the search function. In this way, these functions and their related metrics are inseparable. By induction, subfunctions have their related (sub)metrics. The RCs, therefore, define a set of subfunctions for the functions defined by the MSCs. These subfunctions can be associated with the task designation in the hierarchy of objectives.

When a set of subfunctions is carried out in some prescribed manner, we call that a procedure. This represents the implementation of the higher order function in terms of its subfunctions. If this procedure is modeled, the resulting equation or simulation prescribes the relationship of the lower level metrics to the higher level ones. Figure 5 shows some simple examples of this convention.

Although the audit trail only relates two levels of functions and metrics, further examination of the nature of the requirements reveals more relationships. The twenty-two stressing cases are clustered into three groups. One dealing with area ASW in peacetime, low conflict, or crisis situations; one for area ASW in global conventional war (nuclear war has not been included here); and one involving local ASW in war. There is an increasing scope of operations implied in these three groups as well as a correlation with the size or aggregation of forces involved. When the representative func-

FUNCTIONS AND METRICS

•INSEPARABLE

FUNCTION ←——————————→ METRIC

KILL SUB   PROB (KILL SUB)

DETECT, GIVEN CUE   PROB (DET/CUE)

•PROCEDURES PRESCRIBE AGGREGATION EQUATIONS

--SUBFUNCTIONS PRESCRIBE SUBMETRICS

--EQUATION REFLECTS PROCEDURAL COMBINATION OF SUBFUNCTIONS

•EXAMPLE:

FUNCTIONS & PROCEDURES   METRICS & AGGREGATION

DETECT → CLASSIFY → ATTACK $P(K)=P(K/ATTACK) \cdot P(ATTACK/CLASS) \cdot P(CLASS/DET) \cdot P(DET)$

Figure 5. Relationship of functions and metrics.

tions associated with the metrics for these groups are displayed as in Figure 6, a pattern emerges of a hierarchy of objectives, the top of which depends on the relevant group of situations or missions. (The top of each group, of course, supports higher level objectives, but these are beyond the scope of ASW.) This hierarchy represents a decomposition of warfare functions within the mission domain of our architectural approach. A later section will address the command function decomposition within the organizational domain. The system functions that are defined in the resource domain are implementations of the warfare and command functions. These system functions are supported by equipment functions (such as process, store, and display) or personnel functions.

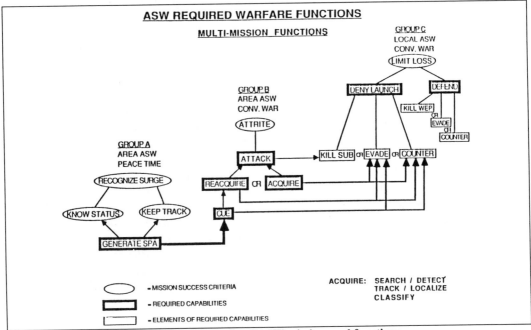

Figure 6. Hierarchy of ASW missions and functions.

MOP/MOE DEFINITIONS

MEASURE OF EFFECTIVENESS*

- PROBABILITY OF SUCCESSFUL ACCOMPLISHMENT OF A FUNCTION
  - FUNCTION: PROCESS RESULTING IN AN OUTCOME OR EVENT
  - ALL PROBABILITIES ARE CONDITIONAL
  - DERIVED FROM MOPs AND LOWER LEVEL OR PRIOR MOE'S

MEASURE OF PERFORMANCE

- NON-PROBABILISTIC PARAMETER OF PERFORMANCE
  - CAUSED BY SYSTEM CONFIGURATION
  - DERIVED FROM PHYSICAL PARAMETERS OR CHARACTERISTICS
  - e.g. GAIN/LOSS, THROUGHPUT, POWER, CAPACITY (#TRACKS)

PHYSICAL PARAMETERS/CHARACTERISTICS

- TANGIBLE ASPECTS OF SYSTEM
  - LENGTH, WEIGHT, CHEMICAL DENSITY
  - NUMBER OF CHANNELS
  - COLOR

* MOFE, MO(P)E, MO(S)E, DEPENDING ON LEVEL

Figure 7. General definition of MOEs and MOPs.

Some of the factors in the audit trail equations are not stated in terms of probabilities. Some of these can be recouched in probabilistic form but some cannot. The latter often represent characteristics, such as speed, that are normally not event-oriented so probabilities are not appropriate metrics. Others, such as throughput, may involve events, but do not represent success or failure, at least not at that level of aggregation. This suggests a dichotomy of metrics into two classes, one of which is defined by the probability of successful accomplishment of a function.

Measures of Effectiveness

Following this suggestion leads to the definitions of measures of effectiveness shown in Figure 7. This way of differentiating among MOEs and MOPs is meant to emphasize the mission orientation of the MOE. In fact, in this approach, the mission (or function) defines the MOE. The MOP class then provides for the collection of metrics, such as gains, rates, capacities, and delays, that are not probabilities of successful outcomes of functions. A third class of metrics represent physically measurable parameters. MOPs are the consequence of a configuration of physical elements. MOEs are the result of a combination of physical parameters, MOPs or other lower level or prior MOEs. [2]

The way that various MOEs are related was implied by the concept of the audit trail, above. In mathematical terms, the MSCs are the dependent variables and they are functions of the independent variables we call RCs. But MSCs and RCs can both be defined in the form of MOEs at the two levels of analysis implied by the audit trail. The analogous relationship can be extended to all levels of a hierarchy of objectives. Then, an MOE defined by an objective function at an upper level is a dependent variable and is a mathematical function of the MOEs defined by objective functions at a lower level. MOE's at the upper level are relevant to the force or system, whose function is defined by that objective. MOEs at the lower level are defined for the (possibly smaller) system designated to perform the tasks or subfunctions. Furthermore, the MOE at the upper level is conditioned on the accomplishment of the lower level functions through the audit trail equation.

This leads to the realization that an important consideration in understanding the signifi-

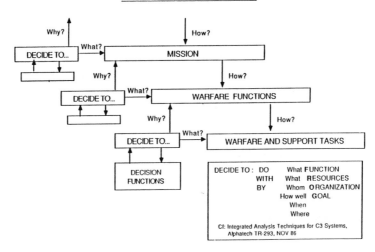

Figure 8.

cance of this approach is the notion that all probabilities are conditional. They are all conditioned on the set (of events or conditions), which is the support for the probability density function. For example, the probability of detection is only defined for the condition that there is a target in the search area. (Otherwise, the event is a false alarm. This is an important complementary measure of effectiveness which supports the objective [or constraint] of not wasting resources.) This notion of conditioning also relates to the hierarchy of objectives. The mission context is what establishes the conditioning on the highest level MOE and function. For example, what is the probability of detection, given the peacetime mission in the Norwegian Sea in summer conditions? When the accomplishment of a function depends on the accomplishment of its subfunctions, the event must occur in the intersection of the subevents. For example, killing an enemy submarine involves detecting, classifying, localizing, attacking AND inflicting lethal damage, given there is an enemy submarine within prosecution range. Then the definition of conditional probability can be used to write the MOEs as chains of conditional probabilities, although analysts are usually guilty of improper notation. A proper audit of the conditional events may be critical to the analysis. In complex analyses involving concurrent or complementary objectives, the calculus of

conditional sets and conditional probabilities that has been devised by (Goodman) and (Calabrese) will be needed to trace these relationships.

The aggregation of probabilities is not a new idea. It is very common in the assessment of operational effectiveness. But the hierarchy of objectives provides a way of formalizing the approach. The role of functions in the hierarchy will provide the setting for examining the effectiveness of C^2 functions.

Role of Decision Making

The key to the definition of C^2 MOEs is the role that decision making has as a function in the hierarchy of objectives. It is a function that occurs at every level of the hierarchy. It is the function that must occur if any other function is to be initiated. In other words, it enables all other functions. Figure 8 shows this concept in an adaptation of Rahmatian's symbology from Figure 2. "What" function to perform is determined by the "Decide" function at that level. "Why" it needs to be performed is established by the higher level objective(s) and is the motivation for the decision. "How" has the same meaning as before. At each level, the "Decide" function is further decomposed into the "Decision Functions" that are needed to make those decisions. We have called them command functions for obvious reasons and they too will play a role in the definition of C² MOEs as will be shown. The box in the lower right corner of Figure 8 acknowledges that this approach bears a strong similarity to and derivation from the Alphatech TR-293 referenced in (Wohl, 1986).

An example of a functional decomposition that incorporates the "Decide" function is shown in Figure 9. Starting with the objective to limit losses, it follows that we can accomplish that by countering the weapon or the enemy platform. Supporting functions are surveillance and cueing. The large "Decide" diamond represents the choice among these and when to execute them. The next level of decomposition is shown as ovals within the four boxes representing the first tier functions.

The smaller diamonds represent a choice among the lower level functions, such as whether to attack or evade the platform or both. This process can be extended ad infinitum. The branching paths among the diamonds represents a hierarchy of decisions that are involved in controlling the activity of the organization and resources whose objective is to limit losses.

There is also a decomposition of the "Decide" function that supports the decision making at each level. Figure 10 depicts a two-tier decomposition of "Command Functions." These functions are related to the functions normally found in any concept model of decision making, such as the Lawson model (Lawson) or SHOR paradigm (Wohl, 1981), et al. We often refer to it as POA&E, for Plan, Observe, Assess, and Execute. The basic structure is intended to be a refinement of the HEAT whirlygig model. A more detailed description of the functions to the third tier is another product of the methodology development mentioned above. The POA&E is shown here in order to highlight two subfunctions, in particular, which are pivotal elements in the decision outcome. The culmination of the decision cycle is the recognition of the appropriate current COA (course of action) and the assignment of available resources to carry it out. These are the top two subfunctions of Execute.

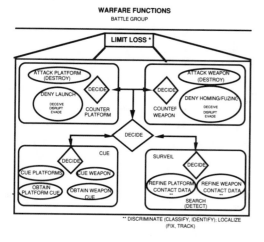

Figure 9. Warfare function decomposition with Decide.

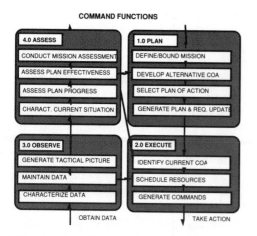

Figure 10. Command function decomposition.

A CANONIC FUNCTION - BASED C2 MOE

PROBABILITY OF DECIDING TO

INITIATE FUNCTION
WITH RESOURCES

(AT A TIME TO EXCEED THE GOAL)

GIVEN

AUTHORITY TO INITIATE FUNCTIONS
AUTHORITY TO CONTROL RESOURCES
INFORMATION NEEDED TO MAKE DECISION
RESOURCE AVAILABLE TO PERFORM FUNCTION

(SITUATION CALLS FOR THE DECISION)

Figure 11. Canonic definition of C² MOE.

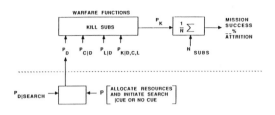

AGGREGATION OF WARFARE METRICS
(MOE's)

IMPACT OF COMMAND FUNCTION/METRIC

**Figure 12. Impact of C² MOE on
warfare analysis.**

C² Measures of Effectiveness

Following the definition of MOE established earlier, the MOE for decision making must involve the probability of making a decision. Figure 11 presents a canonic form for an MOE that reflects the essence of the outcome of the decision process, which is to initiate a function with resources. The necessary conditions for making the decision are also used to "condition" the MOE. Authority is required for each of the elements of the decision: the function initiation and the resource allocation. In addition, resources must be available and information is needed to recognize that a situation exists which calls for the initiation of the function and other information is needed to know which resources are available to assign to the task. Finally, the MOE is conditioned on whether the mission context (current situation) actually calls for the function as being appropriate. This last condition accommodates the concept of the "accuracy" of the decision, which is often mentioned as a C² metric.

Another classical C² metric is timeliness. This is represented by the requirement to initiate the function in time to meet or exceed the goal. The MOE for the mission oriented function will be dependent on and conditioned on the time that it is initiated. If the time of this initiation is within the "window of opportunity," we say that the decision was timely. In fact, the probability of accomplishing a function is a variable function of time. This is also true of the

probability of making a decision. Since the mission MOE is dependent on the decision MOE, their time factors will also be related. Future work is intended to address this relationship.

A third oft-mentioned C² metric is completeness. No aspect of the proposed C² MOE addresses this. I am not sure that completeness is a necessary metric. It usually applies to the information available to the decision maker. The issue here is sufficiency of information, not absolute completeness. Completeness may be an MOP by the definition above and may be useful in assessing equipment performance.

Impact of C² MOE on Analysis

Figures 12 through 14 provide an example of how the C² MOE fits into an analysis. In Figure 12, the normal Probability of Kill is calculated as the product of the conditional probabilities of detection, classification, localization, AND kill. When this is aggregated over all the targets, the mission success for attrition is realized. The probability of detection, in addition to being conditioned on the presence of a target, as was mentioned earlier, is also conditioned on the initiation of the search function. The probability of initiating and allocating resources to search is the C² MOE in this case. It is shown to be conditioned on whether or not a cue has been received. This reflects the difference in likelihood of a decision maker initiating a search, depending on whether there is information available to limit the search area. The probability of detection will be the product of the detection probability conditioned on the search and the search decision probability. Note that the conditional detection probability is also dependent on the availability of cueing information. This suggests a magnification of the effect of a cue since, not only is the detection probability increased due to the smaller area to be searched, the decision probability is also higher.

In Figure 13, the classical role of the sonar equation in determining the probability of detection is shown schematically. This also provides a representation of the role of the definitions for MOP and physical parameters. as noted by the small numerals in the figure. Figure 14 is a pictorial representation of the decision making process (POA&E) and its requisite conditions that provide the basis for the

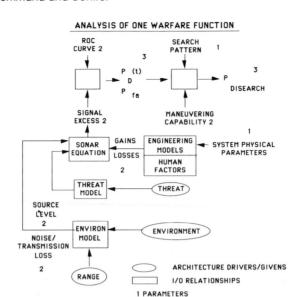

Figure 13. Standard analysis of detection probability.

decision outcome. A variable not mentioned previously is the human factor element. As we know, this will have a major influence on the probability of deciding.

Summary

The final example only shows the role a decision probability has on the outcome of a single mission function. Consider that the concept is intended to apply to the initiation of each and every function involved in the operation. While it may be highly likely that the right decisions will be made when the condi-

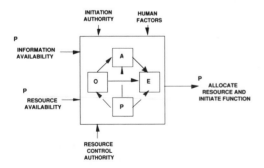

Figure 14. Decision making product relationship.

tions are right and the information and resources are available, the large number of decisions that are involved implies that even small differences from unity will accumulate to deflate the overall effectiveness of the force. The proposed MOE provides the analytical means to account for the effect of "imperfect C^2." It also provides the means to couple the availability of information and resources to the mission outcome. From an organizational perspective, the effect of delegation of authority to initiate action or control resources is explicit and provides the potential for examining dynamic organizational issues.

The Canonic Function-Based C^2 Measure of Effectiveness satisfies the motivational concerns shown in Figure 1. The MOE is embedded in the mission-driven hierarchy of objectives so that the function to be performed provides the context for assessing the effectiveness of the decision making. The MOE is also embedded in the aggregation equation for the operation so that the overall effectiveness is dependent on the outcome of the decision process. Finally, the MOE reflects the purpose of decision making directly, which is to allocate resources to perform functions in a timely manner. The canonic form provides a common definition for all cases and a means for orienting the analyst to the essential conditions to be accounted for, which are information and resource availability and authority to carry out the assigned missions, functions, and tasks.

Acknowledgments

My thanks to Bryson Pennoyer, George Ruptier and Jay Martin of the Naval Ocean Systems Center and Janis Bilmanis of the Naval Surface Weapons Center, White Oak, for their intense efforts in the development of the architectural methodology. And to Vic Monteleon of NOSC for the clue that the purpose of C^2 is to allocate resources. Alex Levis' critique of interdependent MOEs led to clarifying concepts.

Notes

[1]This simple picture masks the complexity of the true interrelationships of functions. A true functional decomposition is not a pure tree and may not be strictly hierarchical. Rather than a tree, a functional structure is more properly represented by a graph. If it is hierarchical, a multiplicity of functions at one level may support or be supported by a multiplicity of functions at another level. Then again, it may not be hierarchical, in that there may be cyclical or mutually supportive purposes, such as sinking submarines in order to protect supply ships which provide logistic support in order to sustain operations to sink submarines. The interdependence or lack of strict ordering of functional relationships does not negate the usefulness of the concept. They will result in simultaneous and higher order equations.

[2]Lower level MOEs are those defined by the next lower level of objective functions in the hierarchy of objectives. Sometimes, objectives at the same level are mutually supportive, such as those in note 1. When this happens, the interdependence of their MOEs must be rectified by causality considerations. Prior MOEs are those associated with objectives that must be accomplished earlier than the objectives that depend on them. In the example of note 1, the sinking of submarines to protect supply ships must occur before the logistic support is effected, which must precede the later sinking of submarines.

References

ASW Top Level Warfare Requirement, Chief of Naval Operations, 1987, Secret.

Calabrese, Philip, "An Algebraic Synthesis of the Foundations of Logic and Probability" *Information Sciences* 42, 1987, pp. 187-237.

Goodman, I.R., "A Measure-Free Approach to Conditioning," *Proceedings of the Third AAAI Workshop on Uncertainty in Artificial Intelligence,* July 10-12, 1987, Seattle, Washington, pp.270-277.

Lawson, Joel S., Jr., "The Art and Science of Military Decisionmaking," *Phalanx*, vol. 15, no. 4, December 1982.

Naval Warfare Publication 1, *Strategic Concepts of the U.S. Navy*, Unclassified.

Rahmatian, Sasan. "The Hierarchy of Objectives: Toward an Integrating Construct in Systems Science," *Systems Research*, vol. 2, no.3, pp. 237-245, 1985.

Signori, David T., and Starr, Stuart H.,"The Mission Oriented Approach to NATO C2 Planning," *SIGNAL*, September 1987, pp. 119-127.

Wohl, Joseph G., and Tenney, Robert R., "Integrated Analysis Techniques for Command, Control, and Communications Systems," Technical Report TR-293, Burlington, Massachusetts: ALPHATECH, Inc., November 1986.

Wohl, Joseph G.,"Force Management Decision Requirements for Air Force Tactical Command and Control" *IEEE Transactions on Systems, Man, and Cybernetics*, vol. SMC-11 , September 1981, pp. 618-639.

12

Evaluating the Impact of Strategic C³ System Performance

Robert W. Grayson

Analytical Objectives

In order to devise an approach for evaluating any components of the national defense structure, first it is necessary to establish the ultimate objective of that effort. What are we trying to accomplish? It is all too easy, otherwise, to get caught up in the details of simulation and synthesis and find that our product is not really useful. I believe that the ultimate objective of any funded activity in the defense community should be to establish or maintain the capability to wage war whenever and however it is in the national interest to do so. I know this puts me in conflict with some deterrence advocates; but without the capability to wage war successfully, I doubt the likelihood of deterrence. It follows, therefore, that I believe the ultimate objective of evaluations, including C³, is to help prioritize the use of our limited defense budget toward the ultimate capability for war fighting in the national interest. We must prioritize development and procurement programs through which we can establish the necessary capabilities. And, we must prioritize training and exercise programs through which we can assure maximum effectiveness of existing resources.

Given these overall analytical objectives, it is necessary for C³ evaluations to support two categories of comparisons:

• C³ systems, subsystems, and components in the context of overall C³ capabilities, and

• C³ systems in the context of overall strategic force capabilities.

The first of these two types of comparisons is to allow prioritization of potentially diverse C³ elements contributing to the accomplishment of the same or diverse C³ missions. Sensor processing versus command center display systems versus communications systems is a fair example. The second of these two comparisons is to allow prioritization of diverse elements contributing to the same or diverse strategic missions. Sensors versus C³ systems versus weapon delivery systems is a common example.

Accomplishment of these comparisons and prioritizations in a traceable, auditable way implies the need for common, mission-related measures of effectiveness. It follows, also, that the measures used must be understandable and useful to a range of people including war planners, C³ planners, engineers, budgeteers, programmers, and targeteers.

Strategic Evaluation Context

It is prudent to recognize that the evaluation of strategic mission success (i.e., projected outcomes for the various levels of potential nuclear conflict) offers a unique context for the analyst. This is evidenced primarily in two ways. First, in contrast to tactical assessments, the evaluation of strategic effectiveness is dominated by weapon considerations. Within the bounds of the scenarios normally considered, the contributions of the engineering of the delivery systems and the physics of the weapons' effects tend to dwarf the contributions of C³ systems. Similarly, evaluations of the total costs involved tend to be dominated by the costs of the weapons and weapon delivery systems. Second, in contrast to evaluations within the Strategic Defense Initiative (SDI) program for example, there is a long-standing precedence in the way potential strategic mission effectiveness is calculated and priorities derived.

Because the impact and cost of C^3 system performance were recognized, or more likely believed, to be so relatively small, strategic analyses were conducted for many years with virtually no consideration given to them. There was a common acceptance of a nebulous "force multiplier" effect attributable to C^3. However, the tendency was to assume perfect C^3 as a base line, thus converting any less-than-perfect C^3 performance, which is likely, into a "force demultiplier."

C^3 analysts attempting seriously to prioritize their systems in the context of overall strategic force capabilities have found themselves in the role of relative newcomers to the strategic force arena. They have been forced to accept and integrate with an analytical process of longstanding. Although new and different methodologies for common mission-related measures of strategic effectiveness have been advanced in the C^3 community, there has been little success in convincing the strategic force community to change. The pragmatic problem, therefore, is to devise a way to integrate or assimilate measures of C^3 effectiveness into the established measures of strategic effectiveness.

Integration of Measures

Strategic force evaluations are based almost exclusively on expected levels of damage potentially achievable against enemy war fighting, war supporting, and C^3 capabilities and occasionally against selected urban and industrial centers. Although sometimes computed with great specificity, the outcome is more frequently expressed in aggregated terms like those implied in Equations 1 and 2, which are shown below.

$$\text{D.E. } j = 1 - [1 - (SSPK_{ij})]N_{ij} \qquad (1)$$

where D.E.j = Cumulative probability that a designated level of damage is achieved on designated targets of type j.

$SSPK_{ij}$ =Single shot probability that a weapon of type i will achieve the designated level of damage against a target of type j.

N_{ij} =Number of weapons of type i allocated against each target of type j.

$$SSPK_{ij} = (PPLS_i)(PLG_i)(PP_i)(PD_{ij}) \qquad (2)$$

where $PPLS_i$ =Probability of prelaunch survival for weapons of type i.

PLG_i =Probability that weapon delivery systems of type i will launch and fly successfully (does not account for weapon accuracy, only delivery system reliability).

PP_i = Probability that weapons of type i will penetrate the appropriate defenses.

PD_{ij} = Probability that a single detonation of type i, delivered with its associated accuracy, can achieve the designated level of damage against a target of type j.

I prefer to call these the "general strategic equations." In a simplistic fashion, they serially link the uncertainties associated with the delivery of nuclear weapons on target and the uncertainties of the weapon effects themselves. The users of these equations group the uncertainties into larger or smaller aggregations consistent with the needs of each particular analysis. Penetration through several layers of defense might each be entered separately, for example, instead of the single probability (PP) shown in Equation 2. The grouping shown, however, has proved most useful to me, particularly because it correlates with the sequence in which the uncertainties would impact chronologically along the weapon flight path, and is thus intuitively clear. In this sense, it may be convenient to view the equation as representing a series of "barriers" through which the weapons must pass. Extending this analogy in 1980, The MITRE Corporation and the Defense Communications Agency added a parameter representing another barrier; namely, the uncertain likelihood that the forces themselves would receive the order to execute. This is called the Probability of Correct Message Receipt (PCMR) and its inclusion is shown in Equation 3.

$$SSPK_{ij} = (PCMR_i)(PPLS_i)(PLG_i)(PP_i)(PD_{ij})$$
$$(3)$$

where $PCMR_i$ =Probability that the order to execute is received by weapons of type i exactly as sent.

This parameter, PCMR, now serves as the principal measure of strategic communications connectivity throughout the defense community.

The linkage of the PCMR measure to strategic force effectiveness is easily recognized and understood and is increasingly being accepted. It would be convenient if the other contributions of C³ systems could be linked in a like manner; by adding the probabilities of correct or timely decisions, for example. This can be done to the satisfaction of some people. However, the circumstances must be rigidly controlled to make those probabilities meaningful and the coupling of C³ performance to them becomes quite tenuous. I believe the solution lies in the reexamination of the general strategic equations (including, of course, PCMR).

Each of the parameters of uncertainty in Equation 3 (PCMR, PPLS, PLG, PP, PD) must be calculated separately for each set of analytical circumstances of interest (weapon type, threat, scenario, etc.). This outcome is most credible when those parameters are calculated within the constraints of reality. For example:

• Only whole numbers of weapons are assigned to each target; i.e., N constrained to be an integer number.

• Range constraints are honored in assigning weapons to targets.

• Footprint constraints are honored in assigning missile reentry vehicles to targets.

• Only alert weapons, or weapons that can be brought on line in a timely manner, are considered for assignment.

The simplifying assumptions implicit in these equations are 1) that the cumulative probabilities associated with individual weapons are independent of each other, and 2) that the parameters of uncertainty also are independent of each other. The impact of the former is generally lost in the aggregation. The impact of the latter can be minimized by selectively computing a range of values for various conditions.

In calculating the parameters of uncertainty, the analyst has the straightforward mechanism for relating small changes in system or subsystem performance directly to the aggregated force effectiveness. This is true, of course, provided that he can express that coupling. C³ performance cannot be coupled to the parameter PLG. It can be related to PD only if the location of the target (arguably a C³ function or

a C³I function) is incorporated. However, it can be related to PPLS and PP. The value of those two parameters vary according to the time in the conflict at which the strategic response is executed; generally, both decreasing as the response is delayed. C³ systems support the processes of decision and execution and dictate the time at which a decision can be made intelligently and executed efficiently. The timeliness of C³ functions (with associated quality levels) is thus an indirect coupling, but nonetheless one that should be acceptable and understandable to those who must establish program and budget priorities.

The character of C³ support to strategic force operations, of course, is not simply a matter of the timeliness of decisions. It is also a matter of the quality of those decisions. The C³ goal is an informed, intelligent decision, made in a timely manner, consistent with the achievement of national objectives in whatever conflict scenarios emerge.

Based on these considerations, I believe the analytical construct shown in Figure 1 allows complete coupling of C³ to strategic force effectiveness.

This construct shows a command and control model with C³ decision making support (before the fact) and decision execution support (after the fact). The measures of effectiveness shown are discussed separately below. They are all mission-oriented and are expressed in or related to the terms of the general strategic equations.

Appropriateness of response selection is a comparison of the target set selected based on the actual perception of the threat (i.e., based on C³ performance) and the target set that same decision maker would have selected given perfect knowledge of the threat. Note that this is not a measure of the quality of the decision, nor is it a measure of whether or not we will win the war. We do not even know how to define "winning" in a nuclear war. It is rather a measure of the C³ system's ability to give the decision maker the best possible basis for a decision. The "appropriate" decision, thus, is the one most closely matched to the planned achievement of national objectives under the circumstances at hand. Figure 2 characterizes the implications of inappropriate decisions.

Achievement of damage goals is a comparison of the damage expectancy associated with

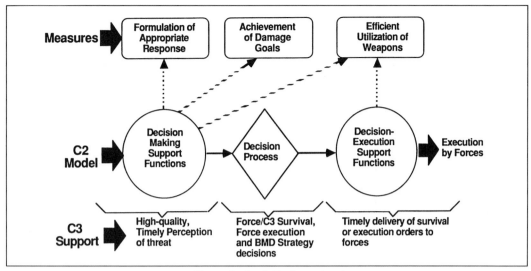

Figure 1. Analytical construct.

the actual execution time and that associated with the execution time that would have accrued with perfect knowledge. As implied earlier, every strategic response has an execution window after which the response becomes increasingly difficult to achieve. The primary causes are the loss of communications connectivity to the forces (i.e., lower PCMR from direct attack and countermeasures), the loss of weapons (i.e., lower PPL from direct attack), and the loss of defense suppression (i.e., lower PP from loss of defense suppression weapons or delays in receipt of the order by those forces).

Efficient utilization of weapons is a comparison of the number of weapons required when execution time is based on the actual perception of the threat and the number required if execution were based on perfect knowledge. Use of this measure implies one of two U.S. capabilities to assign weapons to targets. In the first, war plans might be prepared in advance and stored with the weapons based on two or three values for SSPK reflecting increasingly higher levels of damage incurred by the U.S. C^3 structure and forces. To accommodate the lower values of SSPK, greater numbers of weapons would necessarily be assigned to each target. The decision maker would then choose the appropriate preplanned damage level. In the second, adaptive planning mechanisms

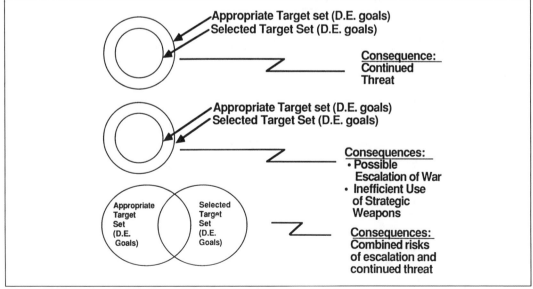

Figure 2. Inappropriate decisions.

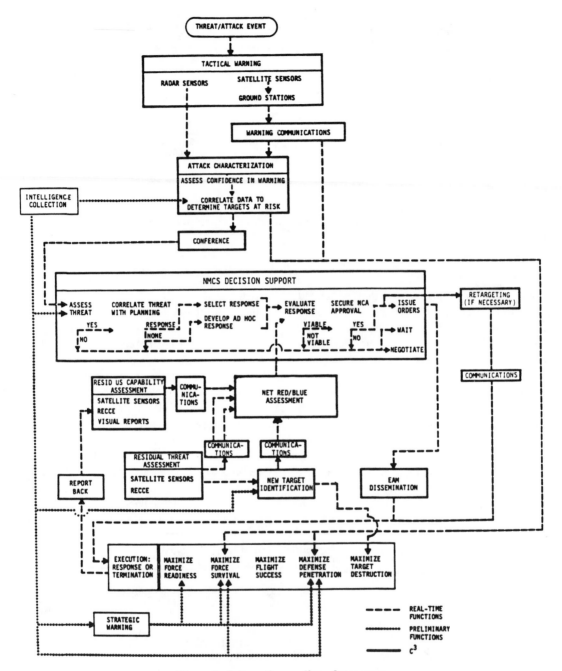

Figure 3. Schematic coupling of measures.

could be developed to allow real time targeting/retargeting at the time the decision is made.

Achievement of damage goals and efficient use of weapons are simply alternative applications of the equations; one application letting D.E. vary, the other letting N vary. The following is a simple example of the alternative uses:

Objective: 0.9 D.E. on each of 20 defended targets

Weapons Assigned: ICBMs

Coupling:

A. SSPK = .6 based on .9 survival and 1.0 penetration, given EAM release $\leq T_1$

B. SSPK = .27 based on .9 survival and .45 penetration, given EAM release $> T_1$

Impact measured by achievement of D.E. goals:

Case 1: EAM released by T_1
   D.E. = $1-[1-SSPK]^N$
   $.9=1-[1-.6]^N$ and
   N=3 per target
Case 2: EAM released after T_1
   D.E. = $1-[1-.27]^3$
   = $1-[.39]$
   = .61

Impact measured by efficient utilization of weapons:

Case 1: EAM released by T_1
   $.9=1-[1-.6]^N$
   N=3 per target (as above)
   60 RVs required
Case 2: EAM released after T_1
   $.9=1-[1-.27]^N$
   N=8
   160 RVs required

The range of coupling of C^3 performance to strategic force decision and execution processes is pictured in Figure 3. Couplings are shown by the dashed lines, real-time functional interfaces, all of which can be expressed in terms of the quality and timeliness of information.

Application of the Methodology

The application of this methodology is pictured in Figure 4.

In the manner already described, the analyst would compare whatever alternatives are available (the right hand side of the figure) against

the reference standard, namely, that which would accrue with perfect knowledge. To do this, the analyst needs the following tools and capabilities.

1. *Subsystem performance models.* There must be detailed models or algorithms with which to calculate the parameters of uncertainty for Equation 3, SSPK. While these are not commonly in use in the offices performing C^3 analyses, they are not hard to find. The Air Force Center for Strategic Analyses (AF/SA), for example, should be able to provide a relatively complete set.

2. *War game model.* A two-sided, two-phase or three-phase model is needed. That is, one in which one side attacks, the second side retaliates and, possibly, the first side strikes again. The model supporting the SIOP-RISOP games (i.e., the U.S. Single Integrated Operations Plan versus the hypothetical Red Integrated Strategic Operations Plan) is the most credible example. Models from the National Test Bed (NTB) being developed for the SDI program would seem to be another credible example, although the specific capabilities of the NTB have not yet been established. Use of either of these models or facilities would likely be difficult and costly for evaluating most C^3 enhancement concepts. Fortunately, however, representative SIOP-RISOP type gaming can be emulated fairly easily.

3. *Decision rules and/or decision makers.* The methodology requires that a decision be

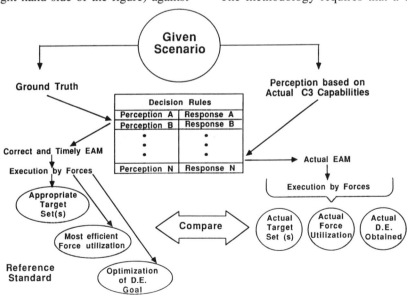

Figure 4. Application of the methodology.

made. The decision need not be "optimum," but must be made in as professional a manner as possible. The best way to do this is to build a set of "if-then" decision rules to serve as guides for response selection. These rules advise the decision maker that for each perception of the situation and threat, a given response is most appropriate, based on preplanning and current guidance. As more decision making role players exercise the model and critique the rules, the more realistic they would become. As this happens, the results from using the methodology obviously would become more credible.

4. *Test driver.* Finally, a test driver model is needed to synthesize the flow of information into the decision process. This can be as simple as a script or as complex as a computer generating outputs from individual sensors or data fusion facilities.

The modeling implied in Figure 4 may be aggregated, using representative but stylized threats, weapon sets, or target bases. It can be event-driven, clock-driven, on-line or using look-up tables. However, for credibility, the analyst must make a conscious effort to be realistic in the kinds of decisions that would be made.

Summary

The methodology described in this paper is much less sophisticated than most operations research professionals would prefer. It is purely pragmatic, based on hard experience (and some success) in making the needs of the C³ community heard and considered in the DOD programming and budgeting arenas.

The analytical processes associated with this methodology are cumbersome. They require literally that the analysts synthesize the outcome of an entire war. However, that is where the ultimate measure of effectiveness lies; namely, in the projected potential capability to wage war in the national interest. And, I believe that adequate war game models exist and can be emulated.

The computational processes associated with this methodology are potentially time-consuming. However, one impact of aggregation is to mitigate this fact.

In summary, the measures generated using this methodology directly relate to the business of national defense and compare directly to the measures commonly used throughout the defense community. The conclusions as to C³ program or budget priorities may not always be accepted. However, the measures used should be understood and acceptable to programmers, war planners, engineers, operators, budgeteers, and targeteers as well as C³ planners.

When the old school general says, "How does C³ make things to boom?," you can say, "I'll show you."

13

Army Maneuver Planning: A Procedural Reasoning Approach

Gregory Stachnick and Jeffrey M. Abram

Introduction

Sun Tzu observed that "a victorious army wins its victories before seeking battle." The goal of the AirLand Battle Management (ALBM) program is to assist U.S. Army planning staffs by providing a collection of cooperating expert systems to support maneuver and fire support planning at both corps and division levels. These planning and decision aids will enhance the staffs' ability to consider more alternatives and evaluate them more rigorously than currently possible. The focus of this paper is one of the ALBM expert systems—the corps level version of the Maneuver Planning Expert System (MOVES).

MOVES is an interactive, knowledge-based, expert system that generates and evaluates courses of action (COAs) for an army corps. The inputs to the system are a mission from echelons above corps, guidance from the corps commander, and data objects containing descriptions of the terrain and weather in the area of operations, the projected enemy situation, and the projected friendly situation; the output is the maneuver portion of an operation order and the COA sketch. The problem, then, is to generate and evaluate multiple COAs that each accomplish the given mission and conform to the commander's guidance for the situation at hand, eventually selecting a single COA from which an operation order (English text and a graphic overlay to a map) is produced. One additional requirement on the system is that it be capable of supporting mixed initiative, i.e., the user controls the degree of autonomy given to the system by inputting in advance as much or as little of a COA as he desires, by modifying decisions made by the system, and by exercising some control over the order in which reasoning processes are performed.

To design and build a system that can solve the problem described above, it was necessary to study and answer the following fundamental questions:

1. How does the U.S. Army plan today?

2. How does one engineer software to perform maneuver planning?

3. What software technology should be used to capture the maneuver planning process?

4. How does one represent a maneuver plan in machine understandable form?

These four questions are answered in the remainder of this paper.

The Manual Maneuver Planning Process

MOVES is an expert system, and as such, its reasoning processes are modeled after those used by human maneuver planning experts. As a starting point, MOVES is modeled after the doctrinal planning process that is described in various Army field manuals [1, 2,] and taught at the Command and General Staff College (CGSC) at Fort Leavenworth, Kansas. This process, called the command estimate, is depicted in Figure 1.

The process is initiated by the receipt of a mission. The mission is analyzed to generate a list of tasks to be performed and a mission statement. The commander can then provide the staff with guidance in areas such as courses of action (COAs) that should or should not be considered, the number of COAs that should be generated, where risk may be acceptable, and command and control arrangements. The staff then gathers the data that will be needed to support the planning process and makes assump-

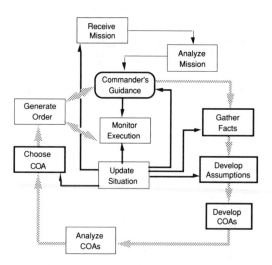

Figure 1. The manual maneuver planning process.

tions, where necessary, as substitutes for unknown critical facts.

The steps performed to this point in the process have set the stage for the first major reasoning task: Develop COAs. In this step, which is performed once for each COA to be generated, a skeletal COA is produced. (The COA is fleshed out further during the Analyze COAs step, which is described subsequently.) The subtasks performed here are:

1. Analyze relative combat power—compare enemy and friendly combat power to infer which types of operations are feasible,

2. Develop scheme of maneuver—choose missions for subordinate units, determine how much force is required for each of these missions, allocate units to missions, and array them on the battlefield,

3. Determine command and control means—allocate major subordinate command (MSC) headquarters to the arrayed forces and determine control measures such as MSC boundaries, phase lines, objectives and assembly areas,

4. Develop COA statements and sketches—produce written and pictorial descriptions of the COA.

In analyzing COAs, each COA is evaluated via war gaming to refine or modify the COA based on the expected outcome, determine the required time sequencing of the various tasks in the COA, estimate the resources required to support the COA, and identify advantages and disadvantages of the COA. This step in the process is highly iterative; war gaming often proceeds until a weakness in the COA is discovered, then the COA is modified using reasoning processes similar to those employed in the Develop COA step, then further war gaming occurs, etc. After all of the COAs have been analyzed, they are compared with one another using criteria identified by the commander, to identify the COA that will be recommended to the commander.

The commander chooses one of the COAs proposed by the staff or asks for modifications, the COA is expanded into an operation plan or order, and the plan is monitored as it is executed. The last task in Figure 1, Update Situation, is a continuous process to collect information from the field to support the information requirements of the command estimate.

The Moves Planning Process

The reasoning process employed within MOVES is shown in Figure 2. First, an intelligent user interface provides an environment for the corps commander or his operations officer to perform the mission analysis and provide guidance to the system, the results of which are represented internally as tasks, guidance, and constraints. Mission and guidance components that specify portions of the solution (resulting plan) are instantiated into the plan object, which is the machine representation of the COA and which is described later in this paper. The initial fragment of the plan object is expanded with an island growing technique to

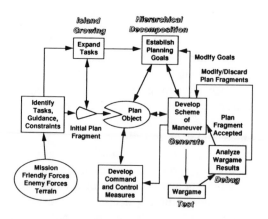

Figure 2. The MOVES planning process.

identify any additional mandatory tasks that must be incorporated into the plan. The resulting plan fragment becomes the common seed from which all COAs will be grown.

The system then hierarchically decomposes the mission and intent into sets of requirements for the principal elements of the battle: deep, security, main, reserve, and rear battle. The different ways of decomposing the overall mission into the goals to be achieved per battle element, each decomposition leading to distinct guidelines for the COA generation process, provides one of the mechanisms in MOVES for producing multiple COAs.

After constructing the initial planning goals, the remainder of the MOVES planning process is an iterative constructive approach consisting of three major subprocess components: geneate, test, and debug.

Problem constraints are employed to conduct a controlled search through plan generation space to construct a plan fragment containing a proposed scheme of maneuver for a selected component of the battlefield. This plan fragment is then evaluated with a heuristic or an analytical war game. If the results of the war game are unfavorable, an attempt is made to debug the current plan fragment to compensate for unexpected shortfalls identified by the evaluation. If the fragment cannot be debugged, then it is not the right solution within the current planning context. It is discarded and a new fragment is generated. For example, there might not be enough force left when the reserve battle is planned to perform a counterattack. Once a satisfactory plan fragment for the battle element has been found, this process iterates over the remaining battle elements.

After completing the development of a scheme of maneuver, command and control arrangements and control measures are generated. This means that the software reasons over the developed force array, hierarchy of goals (tasks), and the terrain efforts, to define the most appropriate organization for the subordinate commands. The results include task organization, unit boundaries, phase lines for control of movement, and assembly areas for uncommitted units with preplanned missions (tasks). Multiple COAs for the same force lay down can be generated by considering alternative means of control. A good example of this would be to generate one COA that has the

corps retain centralized control of the covering force battle and a second COA that decentralizes the control of covering force units to the units in the main battle.

Capturing the Maneuver Planning Process

Because the maneuver planning process is highly procedural in nature, the underlying technology that is used in MOVES to capture the process is procedural reasoning [3]. In the following subsections, we describe the Procedural Reasoning System (PRS) that was developed at Advanced Decision Systems in support of the MOVES development effort, then we discuss how PRS is used to represent maneuver planning knowledge.

The Procedural Reasoning System

PRS provides a method of encoding procedural logic in a visual representation that is understandable and modifiable by domain experts (e.g., Army staff officers), and compilable into efficiently executable code. It supports a goal-directed method of programming that encourages a modular, hierarchical decomposition of goals and subgoals. PRS was inspired by the work of Georgeff and Lansky [4] on plan representation and execution.

PRS combines features of several programming paradigms. As in PROLOG, it encodes inferencing strategies, is goal directed, and features backtracking and retraction of data changes. As in Ada, it has declarative semantics, packaging, and exception handling. As in LISP, it is a symbolic language and facilitates rapid prototyping. As in rule-based systems, PRS also provides a frame system for data representation. Unlike any of these other languages, PRS represents a program visually and includes a graphical editor for program creation and modification.

The fundamental element of a PRS program is called a procedure. A PRS procedure is developed by a knowledge engineer, or possibly by a trained domain expert, and can then be augmented by a software engineer to add declarations and data flow specifications that enable the procedure to be compiled into modular, efficiently executable code. Execution speed of the PRS program is enhanced by performing the bulk of the necessary searching and pattern matching during compilation, at which time the visual representation is translated into executable COMMONLISP code. The visual nature of the language allows a domain expert who is not computer literate to participate more actively in the construction and critiquing of the software than is possible with a standard computer language.

PRS Definitions

A procedure is a network representing a set of instructions for accomplishing a specified goal. As seen in the sample procedure in Figure 3 (which is a screen dump of an actual PRS computer display), boxes represent actions or goals, circles represent states, and directed arcs represent the reasoning flow.

There are a number of different actions that can be used along the arcs of a procedure:

1. Achieve ! effect-name parameters—Find a match for the specified effect in the achievement base. If no match is found, execute each procedure that can achieve the effect until one

succeeds or all fail.

2. Achieve (no test) - !! effect-name parameters—Execute each procedure that can achieve the effect until one succeeds or all fail.

3. Test - ? effect-name parameters—Find a match for the specified effect in the achievement base. If no match is found, backtrack.

4. Perform - @ procedure-name parameters—Execute the specified procedure directly.

There are also actions for asserting affects and for evaluating LISP expressions.

There are a number of different node types that are supported within PRS. A start node is the normal entry point for a procedure. End and Break nodes represent successful completion of a procedure, the Break node allowing reentry to a procedure during backtracking. Raise and Exception nodes allow abnormal exit from a procedure, similar to raising exceptions in Ada.

Executing A Procedure

Any path through the network from the Start node to any End node represents a possible

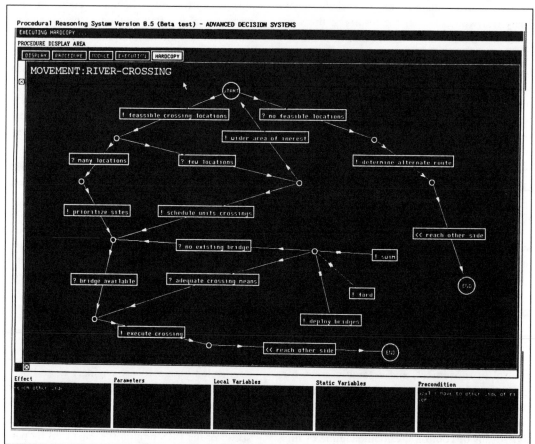

Figure 3.   A sample PRS procedure.

successful execution of the procedure and is called a line of reasoning. A procedure is searched in an ordered depth first, left first manner, thus the order of search is inherent in the topological structure of the procedure.

Backtracking occurs whenever an action fails or a goal fails to be achieved. Execution backs up and attempts the next untried branch. Automatic retraction of assertions and assignments occur simultaneously with backtracking, thus, only actions along a successful path or line of reasoning affect the final results.

To illustrate the execution of a procedure, we now step through the example that was shown in Figure 3. Assume that the procedure, named RIVER-CROSSING, has been invoked either by name or by requested achievement of the effect named "reach other side." Before execution begins, the precondition "crossing type = unopposed" is checked, and only if the precondition is true will the procedure be executed. One can have several procedures capable of achieving the same effect. Some can be general purpose, others can be specific to certain situations. The precondition is one method of controlling which procedures can be considered in a given situation. Assume that the precondition is satisfied.

The first arc executed from the Start node is an achieve action to find all feasible crossing locations, which fails only if there are no feasible locations. If it succeeds, the next arc executed is a test to see if there are many locations to choose from. Suppose this test fails. The search backtracks to the previous node and executes the other outgoing branch, which is a test that there are only a few locations to consider. The system next attempts to achieve the effect called "schedule units crossings," to make efficient use of the limited number of available crossing sites. If this arc fails, execution backtracks again and the arc to achieve a wider area of interest is tried next. Assuming success and returning now to the Start node, feasible crossing locations for the broadened area of interest are found. This time, perhaps there are many locations to chose from, and the next arc prioritizes them. If the test for "bridge available" succeeds, a procedure is invoked to execute the crossing, the system asserts the effect "reach other side," and this procedure is exited with successful completion.

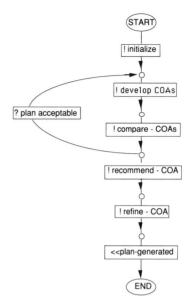

Figure 4. MOVES-TOP procedure.

How PRS Represents the Maneuver Planning Process

The maneuver planning process is encoded within a hierarchy of PRS procedures. High-level procedures encode the knowledge required to solve large subcomponents of the planning problem at a high level of abstraction; low-level procedures encode the fine grained knowledge required to solve small, narrowly scoped subproblems.

At the highest level of the MOVES hierarchy is a procedure called MOVES-TOP, shown in Figure 4. This procedure captures the entire maneuver planning process at a very high level of abstraction. Tracing through the procedure, one sees on the arc labels the steps of developing multiple plans, recommending one COA to the commander, refining the COA to respond to feedback from the commander and expanding the COA into a full plan. The plan is then ready to be processed by a block text generator to produce the operation order.

As one proceeds deeper into the hierarchy, the problem being addressed by each procedure becomes smaller and the reasoning knowledge represented becomes more detailed. At the lowest level, there are primitive procedures whose arcs can invoke LISP code, but cannot invoke subordinate procedures.

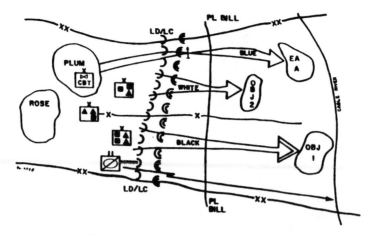

Figure 5. Battlefield geometry component.

Representing the Plan

The results generated by the ALBM planning process must be stored in a representation that is sufficiently complex to capture the complete description of the tactical actions that compose the plan and the relationships that exist between various tasks and physical areas on the battlefield. The plan object consists of four major components:

- task organization
- battlefield geometry
- fire support plan
- tactical action dependency network

Task Organization

The Task Organization (TO) component describes the operational command structure for the plan. It identifies how the various major subordinate commands are structured as a function of the different phases of the plan. It reflects the attach and OPCON (operational control) relationships that have been derived by the planning process to account for shortages or extras in the organic unit compositions.

Battlefield Geometry

The Battlefield Geometry (BG) component of the plan object, represented in Figure 5, contains the information describing terrain-related control measures. This information includes the designated areas of operations for each of the major subordinate commands; assembly areas, attack and counterattack objectives, engagement areas, and axes of advance.

Certain coordination information also is included in this schematic in the form of phase lines. A particular friendly or enemy unit crossing a phase line can be the trigger for termination of some current activity and initiation of the next. The BG object also contains information describing how the various battle elements (security, main deep, rear, reserve) are spatially defined for the course of action. Part of this component is filled in with inputs provided by echelons above corps, for example, the corps area of operations and perhaps the Forward Edge of Battle Area (FEBA) or covering force coordinating points. The remainder of the information is either provided by the system user or is determined as part of the maneuver planning process.

Fire Support Plan

The fire support component of the plan object describes which fire support assets are retained under corps control and which assets are assigned to the major subordinate commands. The specific sets of fire support tasks that support the maneuver missions are incorporated into the tactical action dependency network, described below.

Tactical Action Dependency Network

The tactical action dependency network (TADN) is the most important and most complicated of the plan representation components. This component captures the tasks that have been assigned to the corps and its subordinates, and the various dependency relationships

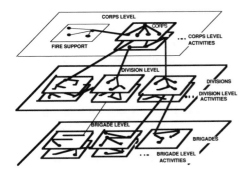

Figure 6. TADN task decomposition.

between tasks that, taken together, describe the sequence of activity making up the plan.

The TADN is a network of task and conditional nodes with the arcs representing hierarchical task decomposition, temporal, and tactical proximity relationships between the nodes. Figure 6 illustrates the plan object from the task decomposition perspective. The TADN begins with a top-level reference node representing the task which corresponds to the given corps mission. The network is hierarchically organized according to the decomposition of tasks into subtasks. Thus, nodes deeper in the structure are more detailed and represent units at lower echelons. In Figure 6, the first decomposition would represent the corps tasks for the deep, close, rear, and reserve battle elements. The specific decomposition of the corps close battle element task could be found in the division level segment of the diagram.

Each TADN node contains a task object and sets of parent links, child links, temporal links, and tactical proximity links. The task object stores basic information about the kind of tactical action that is to be performed on the battlefield. The task object contains the following:

Who—the unit performing the action and

What —the activity to be performed and

What Direct Object—object to which the activity applies and

Where—the location and

When—the time and

Why—the purpose to be achieved and

Why Direct Object—object of the purpose and

How—the specific form of maneuver or another decomposition.

For example, consider the following task:

On order (when) the 208 ACR (who) attacks (what) the 79 Tank Regiment of the 27 Division (what direct object) at objective blue (where) to force deployment (why) of the 9 Tank Regiment of the 27 Division (why-direct-object).

In this example the "how" was not given, indicating that the way in which the mission is accomplished is implicitly described in the task objects of subordinate nodes in the network.

Additional information about the task is also available for determining whether the task is a principal maneuver task or a support task. Some tasks have special invocation and terminations that are not derivable from the temporal relations. For example, when the 208 ACR is pushed back over Phase Line Bill, it transitions from a "defend" mission to a "delay" mission. Crossing the phase line is the termination condition for the "defend" mission.

Whenever the maneuver planner determines that a particular temporal ordering of two tasks is required, the relationship is represented as a temporal link connecting the corresponding task nodes. There are thirteen possible values that describe the way in which tasks can be temporally related, the temporal relationships being based on Allen's theory of time [5]. These relations have been defined as follows:

before (A,B): A ends before B starts

after (A,B): A starts after B ends

overlaps (A,B): A starts before B starts, ends after B starts, and ends before B ends

overlapped-by (A,B): A starts after B starts, ends after B ends

equal (A,B): A and B have same start and end

starts (A,B): B starts when A starts

started-by (A,B): A starts when B starts

ends (A,B): B ends when A ends

ends-by (A,B): A ends when B ends

met-by (A,B): B starts when A ends

during (A,B): A starts after B starts, and A ends before B ends

contains (A,B): A starts before B starts, and A ends after B ends

Allen uses these relationships to describe the possible temporal relationships that can exist between two time intervals. A temporal link between two TADN nodes has a stronger connotation: The link represents a requirement that a particular temporal relationship exists between the two nodes. The information contained in the "when" slots of two nodes can be examined to determine the temporal relationship that exists between the two nodes, but the nodes could be temporally independent and the relationship could be purely coincidental. However, if a temporal link connects the two nodes, then the temporal relationship must be maintained, and changing the time interval associated with one of the nodes may cause the time interval of the other to be adjusted accordingly.

Tactical Proximity

When allocating the command and control measures after forces have been arrayed, it is important to know which tasks in the network strongly influence other tasks in the network. Such tasks could be considered part of a single battle. This relationship between nodes is called tactical proximity. The values for this relation are indicated by tactical proximity links. The acceptable values are:

Tactically-close (A,B):   strong influence

Tactically-distant (A,B):   negligible influence

Tactically-intermediate (A,B):   moderate influence

The concept of "tactically close" does not depend on physical distance only. For example, two defend activities on adjacent avenues of approach may be "tactically distant" if a significant obstacle exists between the avenues. If two tasks are "tactically close," there is strong motivation to assign both tasks to the same major subordinate command.

Conditional Nodes

A single set of tactical actions may not always be adequate to respond to sets of differing enemy actions. In this case the plan representation must have the capability to incorporate a branch. A branch is indicated by a conditional node. A conditional node consists of sets of parent, child, and temporal links, but in addition it has a list of "control forms." Each control form consists of a LISP predicate and an accompanying list of task nodes to be executed if the predicate evaluates to true during plan execution. This predicate corresponds to a test against the world states that result from execution of the plan. For example, the predicate might test whether enemy units move to the north or south at a particular intersection of major roads. If the predicate indicating enemy goes north is true, the nodes corresponding to that branch of the plan are executed as required. These sets of predicates also help to identify requirements for what kind of data needs to be collected about the battle as events unfold.

Conclusion

This paper has discussed the ongoing development of the Maneuver Planning Expert System (MOVES), which is a major component of the AirLand Battle Management program. Both the manual process used in the U.S. Army today and the process model built into the system are highly procedural in nature, leading to the conclusion that the PRS, developed at Advanced Decision Systems, is an appropriate technology with which to implement the MOVES system.

PRS has a number of features that make it an appropriate software development environment for MOVES development:

1. It is designed specifically to address the problem of representing procedural knowledge.

2. Its graphical representation of both domain knowledge and inference control mechanisms make the inherent reasoning processes visual to the software developer and, more importantly, to the computer illiterate domain expert, who can then interact with the developer directly through the code, rather than through the developer's explanation of what the code is supposed to do.

3. The PRS programming paradigm encourages the development of modular, hierarchically structured software. It supports the ability to develop the reasoning knowledge only down to a certain, desired depth, with the logic below that level easily "stubbed" out until it is desired to fill in more low-level knowledge. The PRS development environment supports an incremental approach to software development through rapid prototyping.

4. Much of the search and pattern matching

is performed at compilation time, providing excellent performance at execution time, unlike most expert system development environments.

PRS is used to represent the maneuver planning reasoning processes, but is not used to represent the plan that is dynamically generated by the MOVES system. A complex plan object has been developed to represent the actions that make up a plan, the various dependency relationships in effect between these actions, and the measures in place to control the plan.

Over the remainder of the project, MOVES will evolve as we continue to expand the knowledge encoded in the system in the form of PRS procedures, adding more breadth to the types of missions for which the system can generate plans, and adding more depth to the knowledge already captured.

Acknowledgments

The authors wish to express their gratitude for the helpful comments provided by James P. Marsh and James R. Greenwood, Advanced Decision Systems, and by Morton A. Hirschberg, U.S. Army Ballistic Research Laboratory, Aberdeen Proving Grounds, Maryland. We also wish to acknowledge the efforts of our ALBM project coworkers.

References

1. Department of the Army, FM 100-5, *Operations,* 1986.

2. U.S. Army Command and General Staff College, Student Text 100-9, *The Command Estimate,* 1988.

3. Georgeff, Michael P., and Lansky, Amy L., "Procedural Expert Systems," in *Proceedings of the Eight International Joint Conference on Artificial Intelligence,* Karlsruhe, Germany, 1983.

4. Georgeff, Michael P., and Lansky, Amy L., "A System or Reasoning in Dynamic Domains: Fault Diagnosis on the Space Shuttle," Technical Note 375, SRI International, 1986.

5. Allen, James F., "Towards a General Theory of Action and Time," *Artificial Intelligence,* vol. 23, 1984.

*This research was supported by the Defense Advanced Research Projects Agency (DARPA) and the U.S. Army Ballistic Research Laboratory (BRL) under subcontract 7A40B0820M from Lockheed Missiles and Space Company, Inc., Austin Division.

14

Multitarget Tracking and Threat Evaluation in Airborne Surveillance

Thomas Kurien and Adam Caromicoli

1. Introduction

The mission of the E-2C Hawkeye is to conduct airborne surveillance for a Naval battle group. To carry out this mission, the E-2C surveillance system has a suite of on board sensors, communication systems linking it to other platforms in the battle group, on board data processing capabilities, and a crew of operators. The on board sensor suite includes a UHF radar, an identify-friend-or-foe (IFF) system which can interrogate friendly targets, and an electronic support measures (ESM) system which can detect and identify radars on board targets. Target track data available to other platforms in the battle group is accessible to the E-2C over the communication links. Using the sensor data and the track data over the communication links, the surveillance data processing computer must track the position, velocity, and identity of all targets within the surveillance volume. Based on this target track data, the E-2C operator must evaluate a measure of the threat faced by the battle group.

During the past few years, the amount of surveillance data accessible to the E-2C has increased severalfold. This has been a result of both the extended coverage regions of the sensors made possible by improvements in sensor and communication systems on board the E-2C, and the vast number of platforms with associated electronic warfare systems (e.g., ESM, ECM, and ECCM) present in modern tactical scenarios. The large amount of surveillance data has burdened the E-2C surveillance operators with monitoring the results of the current data processing algorithms. Such tasks include maintaining track continuity of radar tracks and correlating data from different sensors. Not only does this cause the operator's attention to be consumed by the task of monitoring and correlating the vast number of tracks—a task for which he (or any human) is not really suited, but it also causes him to neglect his primary duty of evaluating threat to the battle group—a task for which he is better suited.

To keep pace with the data-rich environment, a high speed processor (HSP) has been incorporated into the E-2C. This presents an opportunity to incorporate advanced data processing algorithms that can better support the surveillance operators and thereby enhance the mission effectiveness of the E-2C system. The objective of our effort is to develop a prototype of such a data processing algorithm. Notice that the data processing algorithm is intended to serve as a decision aid and not as a replacement for the surveillance operator. We do not foresee existing or near-term algorithm technology replacing a human operator for a function as critical as threat evaluation in tactical warfare.

The objective of this paper is to discuss the overall architecture of the decision aid and to examine in detail the threat evaluation algorithm. Section 2 provides an overview of the architecture of the decision aid. The correlation and tracking, and the identification algorithms are discussed briefly in Section 3. Details of the threat evaluation algorithm are provided in Section 4, and simulation results, which demonstrate the effectiveness of this algorithm, are provided in Section 5.

2. Architecture of the Decision Aid

The first step in the design of a data processing algorithm is to clearly define all a priori knowledge available about the system. For the

E-2C airborne surveillance system, this entails two bodies of knowledge:

1.   Physical laws that model well-understood phenomena such as the dynamics of target motion (e.g., Newton's laws of motion), and the measurements from the E-2C sensors (e.g., electromagnetic propagation and reflection laws for the radar signal). These also include statistical models, which account for the lack of precise knowledge of the dynamics (e.g., effect that wind has on aircraft motion) and measurements (e.g., measurement uncertainty introduced by randomly changing characteristics of the medium through which the radar signal propagates), and geometrical relationships, which account for sensor-target geometry (e.g., Pythagorean theorem).

2.   "Man-made laws" such as doctrines used in tactical warfare (e.g., friendly aircraft approaching an aircraft carrier have to follow flight corridors), "man-made rules" for evaluating threat (a fighter aircraft has tactical air superiority over a bomber), and known facts (e.g., the BADGER is a Soviet bomber carrying a type A tail warning radar, has a maximum speed of B knots, and is equipped with type C antiship missiles that can be launched from a range of up to D nautical miles).

Data processing algorithms designed to utilize these two bodies of knowledge have different characteristics. For the former body of knowledge, algorithms may be designed that are based on a precise mathematical framework and follow a fixed sequence of steps to process the data. Further, they are entirely *numerical* in that these algorithms process numerical data while carrying out dynamic, geometrical, and/or averaging transformations to arrive at the result.

One cannot formulate precise numerical relationships for the latter body of knowledge, because they either do not exist or are not known. Variables associated with this body of knowledge are described in a symbolic or aggregated form. One cannot formulate a fixed sequence of steps to arrive at the desired result; rather, one must process all the available data to derive a recognizable pattern. This form of processing is generally *symbolic* in nature. It should also be noted that humans process data in symbolic form and prefer to interface with symbolic rather than numerical algorithms.

Since the E-2C surveillance problem entails both bodies of knowledge and since the data processing algorithm must interface with the E-2C operator, the design of the data processing algorithm requires both forms of processing. A conceptual architecture for combining the two forms of processing in the decision aid is shown in Figure 1. The figure illustrates a partitioning, with the numerical processing algorithms on one side and the symbolic processing algorithms on the other. The numerical algorithms are constructed based on models and associated parameters of the physical laws. They process the numerical data from sensor reports to evaluate numerical results. Symbolic algorithms utilize rules and facts to process the symbolic data. They evaluate the threat conditions for the battle group. The two forms of processing are linked by a common database that contains intermediate and partial results produced by either form of processing and required by the other. Specifically, the numerical algorithms compute numerical results that enable certain rules to *fire* within the symbolic algorithms, and the symbolic algorithms will request new results to be evaluated by the numerical algorithms. Explanations for generated results are produced by the symbolic algorithms in response to operator queries.

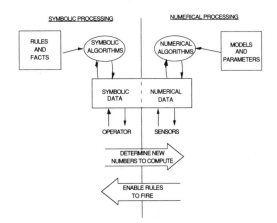

Figure 1. Conceptual architecture of the decision aid.

A functional architecture of the decision aid is shown in Figure 2. The two rectangles drawn with broken lines represent the numerical and symbolic processing partitions discussed above. Specifically, the algorithms for correlation and tracking, and target identification are numerical algorithms; the remaining

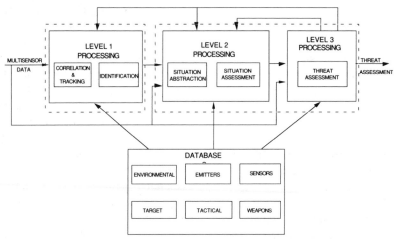

Figure 2. Functional architecture of the decision aid.

algorithms are symbolic. We have chosen titles for the modules (Level 1, 2, and 3 Processing) and the submodules (Situation Abstraction, Situation Assessment, etc.), which are consistent with the lexicon defined by the Fusion Subpanel of the Joint Directors of Laboratories (JDL) C[3] Research and Technology Program [1]. The functions of each of the submodules are as follows: the (Multisensor Multitarget) Tracking and Correlation submodule correlates the sensor reports over time and sensors and evaluates the position and velocity of all targets; targets are identified based on the sensor reports by the Identification submodule; target characteristics relevant for threat evaluation are abstracted by the Situation Abstraction submodule; the Situation Assessment submodule evaluates the hostile target threat and the ability of friendly forces to engage the hostile targets effectively on a one-on-one basis; finally, the Threat Assessment submodule evaluates the collective hostile target threat and the ability of the battle group to engage the enemy effectively.

3. Target Track Correlation and Identification

Both the Correlation and Tracking, and the Identification problems can be modeled fairly precisely. Based on these models, we have developed correlation and tracking, and identification algorithms which follow a fixed sequence of steps to process the sensor data. For reasons discussed in Section 2, we classify these as numerical algorithms. We will discuss

these numerical algorithms briefly in this section.

3.1. Correlation and Tracking Algorithm

The multisensor multitarget correlation and tracking problem has been studied extensively in the past [2]. Some of these algorithms, especially those currently used in operational systems, are based on heuristic rules formulated using intuition and experience from actual surveillance scenarios. Such algorithms work well in specific situations; however, because they are not based on a general model, they fail to handle situations other than those considered during the design.

Other correlation and tracking algorithms are based on principles of statistical estimation theory [3-5]. These approaches define a precise mathematical model for the system, formulate the equivalent mathematical problem, and develop the algorithm to compute the optimal solution. Such approaches work well in most situations, but they generally have large computational requirements. However, carefully designed heuristic rules may be designed and incorporated into the optimal solution that dramatically cut the computational requirements of the optimal algorithm without significant loss in performance. In fact, such an algorithm has been demonstrated in real time using real data [6]. The same algorithm has been incorporated into our decision aid. We will overview the mathematical approach and the heuristics used to develop this multisensor multitarget tracking algorithm here; the reader may refer to [5] for a detailed description of the algorithm.

The correlation and tracking algorithm uses the mathematical framework of Hybrid State Estimation to formulate the solution methodology. A general hybrid state model consists of continuous-valued states and discrete-valued states. Measurements related to the hybrid state are used to evaluate an optimal (minimum-mean-squared-error or maximum-a-posterior) estimate of the hybrid state. Variables in multisensor multitarget tracking can be identified with the generic hybrid model as follows: the position and velocity of all targets constitute the continuous-valued state; indicators for target status (constant velocity model, maneuver model) and sensor report status (report associated with a specific target, report is a false alarm) constitute the discrete-valued state; the noisy measurements of range, angle, etc. from the sensors constitute the measurements.

The optimal estimator for the hybrid state based on the measurements is easy to formulate; however, it is not possible to implement the optimal algorithm because the number of discrete-valued states (referred to as global hypotheses) grows exponentially with time. In order to construct a practical algorithm, all the unlikely global hypotheses must be pruned. The key techniques that we have used to prune unlikely hypotheses are outlined below.

N-scan Approximation: The optimal correlation and tracking algorithm requires that each track postulated for a target be associated with each sensor report since all such associations may be possible. In reality, we know that each target should have only one track corresponding either to an association with the report it generates or to no association if it is not detected. The N-scan approximation waits for N scans before it resolves multiple associations made for a target in a particular scan.

Gating: Gating is a screening technique that prevents the formation of unlikely associations of sensor reports with targets. The gating process consists of constructing a region (gate) around the predicted target position, and selecting only those reports that lie within this region to be associated with the target track. Gating is effective in cutting down the number of unlikely target-to-report associations and is commonly used in tracking algorithms.

Classification: Another powerful screening technique that we have incorporated involves the selection of only a group of targets while forming global hypotheses. The selection is based on the criterion that the target track should have a likelihood greater than a threshold. Targets that satisfy this criterion are referred to as Confirmed targets. The remaining targets are grouped as either Intermediate, Tentative, or Born targets and each group has its own likelihood threshold. This form of grouping is termed Classification.

Clustering: As mentioned earlier, the computational complexity of the correlation and tracking algorithm arises mainly due to the vast number of global hypotheses that may be formed. The number of global hypotheses is an exponential function of the number of target tracks postulated. If target tracks lie within different regions of the surveillance volume, such that no common reports are assigned to them, then obviously there is no need to form global hypotheses across these tracks. Clustering is a technique for grouping target tracks in order to avoid the formation of such global hypotheses.

3.2 Identification Algorithm

The target identification problem also has been studied extensively in the past. Methodologies generally differ with regard to how they model and account for the uncertainty in the sensor measurements [7]. We have adopted a Bayesian method since a Bayesian identification algorithm will be compatible with the correlation and tracking algorithm discussed in subsection 3.1. We will overview the mathematical approach and the heuristics used to develop this multisensor multitarget identification algorithm here; the reader may refer to [8] for a detailed description of the algorithm.

E-2C sensor observables that may be used to identify targets include the types of emitters on board targets reported by the ESM sensor, the speed profiles of targets evaluated by the tracking algorithm, the trajectory profiles of targets evaluated by the tracking algorithm, and the IFF responses from targets reported by the IFF sensor. For each of these sensor observables we have constructed a model that shows its dependence on the identity of the target from which the observable originates, and on the sensor processing characteristics used for evaluating the observable. It would be ideal if all the sensor observables could be related to the target identity at the Type level (Bear, F-14, etc.); however, this is not possible because some of the sensor observables cannot discern targets at the Type level. Examples of such sensor observables are: IFF responses from a target, which can determine only whether a target is friendly or neutral; and speed profile of a target, which can decide only whether a target is a fighter or a bomber. In such cases, we have related the sensor observable to the target identity at two higher (more abstract) levels which correspond to Class (bomber, fighter), and nature (friendly, neutral, hostile).

Based on these models for the sensor observables, we have derived the optimal solution to the multitarget identification problem [8]. It is an extension of the optimal solution to the multitarget correlation and tracking problem and accounts for the following factors:

• the state of each target includes the kinematic state as well as the identity state;

• the association of sensor reports with a target track is aided by the target identity; and

• the set of correlated sensor reports for a particular target may be used to evaluate the optimal estimate of the identity for that target.

We have already indicated that it is not possible to evaluate the optimal solution to the correlation problem. We also have shown some of the heuristics to reduce the computational requirements of the optimal correlation algorithm. Here we will indicate how the target identity can be used to aid the correlation process, and some important heuristics used to reduce the computational requirements of the optimal identification algorithm.

Use of Target Identity for Gating: As indicated earlier, gates represent regions in the measurement space within which sensor reports for postulated target tracks are expected to lie. For the tracking problem, the measurement space is confined to kinematic observables. For the combined tracking and identification problem, this measurement space can be expanded to include the sensor observables related to target identity. Gating of sensor reports based on target identity corresponds to picking only those sensor reports for which the probabilities of the sensor observable conditioned on the target identity is higher than a preselected threshold. The number of sensor reports associated with a target track can be reduced by gating in the combined measurement space rather than just

the kinematic space. This reduction is most pronounced for the case of ESM reports.

Revised Likelihoods for Target Tracks: The state of a target track represents the estimated position, velocity, and identity of the postulated track. The likelihood of the target track is a measure of the probability that the set of correlated sensor reports and dynamic models used to reconstruct the target track state is correct. Target tracks with small likelihoods are likely to be false tracks and can be pruned. For the tracking problem, the track likelihood is evaluated based on kinematic information alone; for the combined tracking and identification problem, the track likelihood also accounts for the target identity and consequently will be more effective in discerning true target tracks.

Overcoming Errors Caused by Incorrect Correlations: The N-scan approximation resolves report associations for a particular scan after processing N subsequent scans of sensor data. It is still possible to make an incorrect association of a sensor report with a target track. An incorrect association will lead to an incorrect evaluation of the target identity. To remove the impact of such incorrect associations, prior to processing the next scan we modify the probability mass estimated for each target identity by a small amount. This enables the identification algorithm to overcome the error in the estimated target identity over the course of processing subsequent scans. The modification of target identity probabilities between scans is analogous to the addition of target dynamic model uncertainty in target tracking.

Multiple Levels of Target Identity: Earlier we indicated that some of the sensor observables are related to the target identity at levels higher than the Type level. We have used Pearl's approach [9] to update the target identity at the lowest (Type) level based on sensor observables related to higher levels. This approach makes the assumption that the sensor observables related to higher levels of target identity are independent of a specific target Type when conditioned on membership in a set that contains that Type.

4. Threat Evaluation

A brute force methodology for evaluating the hostile target threat is to enumerate all tactical scenarios encountered by naval battle groups and assign a threat condition to each. However, the tactical scenarios that naval battle groups will confront today will involve a vast number of targets. Since the number of scenarios is a combinatorial function of the number of targets, their identities, and their relative positions and velocities, it is virtually impossible to enumerate all possible tactical scenarios. Consequently, it will not be possible to use brute force methodology. Perhaps, this also explains why there is no universally accepted procedure for quantitatively or even qualitatively measuring the threat projected by hostile forces against a naval battle group.

One way of avoiding the combinatorial problem is to evaluate the threat hierarchically. Using a hierarchical approach, target threat characteristics are aggregated at several levels. This approach is attractive because it is similar to that used by experienced E-2C surveillance operators. Consequently, they can guide the selection of the number of levels in the hierarchy, along with the grouping and aggregation at each level. A hierarchical approach has the added advantage that explanations for threat conditions evaluated by the decision aid may be provided in the same hierarchical order; such explanations will be easier for the operator to comprehend.

Target correlation and tracking, and identification may be viewed as the first level of processing in the hierarchical threat evaluation procedure. Abstraction of the target characteristics and evaluation of the threat conditions represents the second level in this hierarchical procedure. Finally, the evaluation of the collective threat condition for the battle group represents the third level. As we have indicated earlier, the definition of threat at both Levels 2 and 3 is subjective and so a precise relationship between the threat and the target characteristics may not have any practical significance. Accordingly, we define these relationships based on the lines of reasoning used by experienced operators. For reasons discussed in Section 2, we classify the resulting algorithms as symbolic. We will provide an overview of these symbolic algorithms for Situation Abstraction, Situation Assessment, and Threat Assessment in this section. The reader may refer to [10] for a more detailed discussion of these algorithms.

4.1 Situation Abstraction

The function of the Situation Abstraction submodule is to abstract target characteristics for threat evaluation. These characteristics include not only those evaluated by the Level 1 Processing module such as the target's position, velocity, and identity, but also those associated with the target's warfare capabilities, which are available in the target database. The former characteristics can be parameterized and measured quantitatively (e.g., the hostile target has a speed of 250 knots), whereas the latter characteristics can be parameterized only qualitatively (e.g., the friendly target has a defensive capability of, for example, low). For the purpose of evaluating the threat condition, however, it will be more convenient to transform even the quantitative characteristics to a qualitative form. The reason is that a precise mathematical relationship between the quantitative characteristics and the threat condition is not known. Even if a precise mathematical relationship can be postulated (and such a relationship is likely to be nonlinear), there is little empirical data that can be used to validate such a relationship. At best we might be able to use predictions of this relationship made by an experienced operator. Furthermore, an E-2C operator would, most likely, parameterize all characteristics in a qualitative form.

The characteristics of a friendly and a hostile target that are relevant to threat evaluation are:
- military/economic value of the friendly target;
- category (air or surface) of the friendly target;
- air-to-air firepower of the friendly target (if it is an air target), or the surface-to-air firepower of the friendly target (if it is a surface target);
- location of the friendly target;
- velocity of the friendly target;
- mode in which the friendly target is operating;
- military/economic value of the hostile target;
- air-to-air firepower of the hostile target;
- air-to-surface firepower of the hostile target;
- location of the hostile target;
- velocity of the hostile target; and
- mode in which the hostile target is operating.

The firepower of a target represents both the

offensive and defensive components. The mode (or intent) of a hostile target represents the operating state or condition of the target and is evaluated based on the emitters being used (e.g., if the target has turned on its fire-control radar, it portends the launch of a missile), and the position of the target relative to the other targets.

Since the complexity of the threat evaluation algorithm is exponentially related to the number of targets in the scenario, it is beneficial to group (cluster) targets that act in unison. The selection of targets that will be members of each cluster is based on their interactive characteristics. Specifically, members of each cluster should have small position and velocity separations, a similar nature (friendly or hostile), and the same category (air or surface).

We may aggregate the *intrinsic* characteristics of the individual targets within a cluster since we assume that they act in unison. The intrinsic characteristics of a target include:

• military / economic value,
• air-to-air firepower,
• air-to-surface/surface-to-air firepower, and
• air/surface category.

The aggregation of each of these intrinsic characteristics for the targets in a cluster yield the intrinsic characteristics for the cluster. For example, the aggregation of the military/economic values of each of the targets yields the military/economic value of the cluster.

The grouping and aggregation performed by the Situation Abstraction submodule serve two purposes. First, by grouping the targets into clusters and treating them as single entities in the subsequent analysis, the computational burden is significantly reduced. Secondly, by aggregating the intrinsic characteristics for a single cluster, the synergistic effects of targets acting in unison (such as those of a *tactical unit*) can be captured. For example, two fighter aircraft acting in unison may possess a greater defensive and/or offensive capability than when acting independently.

4.2 Situation Assessment

The function of the Situation Assessment submodule is to evaluate the hostile target threat and the ability of the battle group to engage the targets effectively on a one-on-one basis. To perform this function, the Situation Assessment submodule forms *cluster pairs*.

Each consists of a friendly and a hostile cluster. For each cluster pair, the relative characteristics between the friendly and hostile clusters are evaluated (this represents the aggregation step). These relative characteristics include:

• value-ratio,
• relative air-to-air and air-to-surface capabilities,
• proximity,
• air/surface category of friendly cluster, and
• mode of the hostile cluster.

An example of the evaluation of these characteristics is the proximity of the friendly and hostile clusters. Our definition of the proximity of two clusters accounts for not only the distance separation of the clusters, but also the time to engage, weapon range and the closest point of approach. To understand why this is necessary, consider two cases where the clusters have the same distance separation. In the first case, the clusters are heading directly away from each other, while in the second the targets are headed directly toward one another. In these cases, the proximity would account for the fact that the "separation" in the first case is somewhat smaller in a tactical sense, because their distance separation is decreasing. In addition to the distance separation and the relative velocities of the clusters, the range of the weapons carried by the members of the clusters is included in the evaluation of proximity. This is required because two clusters, that are separated by a significant distance, may still be within the firing range. Finally, the closest point of approach is also included, providing a measure of how close the targets could become if their velocities remain unchanged. A similar aggregation of intrinsic characteristics (i.e., weapon range in this case) and interactive characteristics (i.e. physical separation, closest point of approach and closing velocity in this case) can be used to evaluate the other characteristics listed above.

The next step is to combine the relative characteristics to evaluate a threat condition for each cluster pair. As pointed out above, it may not be possible to derive a precise mathematical formula, and even if we did it is not likely to be meaningful. Therefore, it is more appropriate to relate the characteristics to the threat conditions in a tabular form. The results of the threat evaluation will be qualitative since this is the type of information that operators will provide (i.e., an operator would evaluate the threat in a given scenario as serious rather than a precise number such as 8.357). In addition, it is unlikely that the information required for threat analysis will be provided for the complete list of combinations of the parameters described above. Instead, it is likely that some variables will become important only when other variables take on specific values. For example, the air-to-surface capabilities will be irrelevant in cases where both the friendly and hostile clusters are comprised only of aircraft, while the value and military capabilities of two clusters may be unimportant if the clusters are far apart and not headed toward each other.

Benefits from the aggregation of threat characteristics in the Situation Assessment submodule are the same as in the Situation Abstraction submodule. Specifically, as indicated earlier, the relationship between the cluster characteristics and threat condition must be summarized in tabular form. If the table is to have a realistic number of entries, the number of independent variables must be restricted since the number of entries will increase exponentially with the number of independent variables. The aggregation performed by the Situation Assessment submodule reduces the number of independent variables used to determine the threat condition from 12 to 5 (for example, the proximity evaluation aggregates four parameters into one), thereby making a tabular approach feasible. An additional benefit that results from aggregation is related to the generation of explanations for evaluated threat conditions: It is unlikely that a threat condition evaluated for a pair of clusters based on twelve characteristics will be quickly and easily understood by an operator. Once the aggregation is performed, the explanation can be provided by the assessment module in a hierarchical fashion. This will also enable the operator to ask for explanations at any desired level and detail.

4.3 Threat Assessment

The function of the Threat Assessment submodule is to evaluate collective hostile target threat, and the ability of the friendly targets to engage the hostile targets effectively. The collective hostile target threat represents the overall threat faced by each friendly cluster, based on all hostile clusters which might interact with the friendly cluster. A cluster is defined to interact with another cluster if either of them imposes on the other a cluster pair threat greater than a threshold.

Evaluation of the collective hostile target threat is based on cluster pair threats and the time to engage evaluated by the Situation Assessment submodule. This evaluation procedure also includes the steps of grouping and aggregation of threat characteristics. The grouping step combines all clusters that influence one another's interactions to form what we call super clusters. The threat condition for each friendly cluster belonging to a particular super cluster is evaluated based solely on the clusters belonging to the same super cluster. The aggregation step combines the threat parameters associated with cluster pairs included in a super cluster to evaluate the overall threat for each cluster. Since each cluster within a super cluster has an influence on the remaining clusters, the overall threat for a friendly cluster is related to the threat parameters of all clusters included in the super cluster. The time for each attacking cluster to reach the region from which it can engage the attacked cluster can be calculated based on the relative kinematics of the two clusters. Based on these times, the sequence of engagements within the super cluster can be predicted. Each such engagement increases the threat level to the attacked cluster; at the same time it reduces the firepower capability of the attacking cluster. The overall threat to each friendly cluster is evaluated by stepping through all the feasible engagements and recursively updating the firepower capabilities and the threat levels of the clusters involved in the engagement

5. Simulation Results ─────────────

The Threat Evaluation algorithm discussed in Section 4 has been implemented in PROLOG on a Micro Vax II graphics work station. We refer to this implementation as the Threat Evaluation Module (TEM). The preliminary implementation of TEM is deterministic in that we assume that target tracks and identification are known with probability one. This allows us to "short circuit" the correlation and tracking, and the identification algorithms (discussed briefly in Section 3), and demonstrate the performance of the threat evaluation algorithm.

We plan to implement the probabilistic version of TEM using the same hierarchical approach. This probabilistic version will be integrated with the correlation and tracking algorithms in the prototype E-2C decision aid.

The deterministic implementation of TEM required 31 PROLOG rules and 89 PROLOG facts. In addition, four routines which carry out numerical computations are coded in FORTRAN. We predict that for the probabilistic version of TEM, the number of facts will increase to about 250, and the number of rules to about 50.

We have developed a simulation package to drive the E-2C tactical decision aid. We refer to this package as REDGEN (Realistic Data Generator). It has the capability to generate sensor reports for fairly realistic scenarios that might be encountered. We have used REDGEN to simulate the following scenario to demonstrate the important capabilities of the Threat Evaluation algorithm.

The scenario comprises several friendly and hostile targets. The friendly targets forming the battle group consists of a single aircraft carrier, two destroyers, six combat air patrol (CAP) aircraft (F-14s), and the E-2C. The hostile targets consist of three bombers (Badgers) and six escort fighters (MiG-27s). Figure 3 shows the initial location and a track history of all targets for the duration of the scenario. For

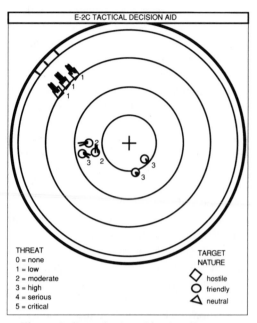

Figure 4. Stage 1 - Attacking hostile targets, CAP aircraft not assigned.

the sake of clarity, we have labeled clusters of targets as opposed to the individual targets; however, we have shown the individual target tracks for the entire duration of the scenario. The racetrack trajectory at the center of the figure represents the trajectory of the E-2C. The concentric circles are spaced 50 nautical miles apart.

The scenario progresses through four stages corresponding to different threat levels. Threats evaluated by the decision aid for conditions that are representative of these four stages are shown in Figures. 4 through 7. In each of these figures, the ESM reports are shown using radial lines in the outer band of the figure. The target tracks for the recent past are shown by short line segments. The nature of the target clusters (friendly, hostile,) is represented by geometric symbols surrounding the target cluster. The meaning of the symbol is indicated in the Target Nature legend. Finally, the current threat posed against a target cluster is shown by a number placed next to the geometric symbol. The interpretation of this number is indicated in the Threat legend. (On the VR-290 color graphics terminal used to display the results of the decision aid, the threat levels are indicated using a color code.)

Figure 4 shows the first stage of the scenario. The hostile targets are assumed to have entered

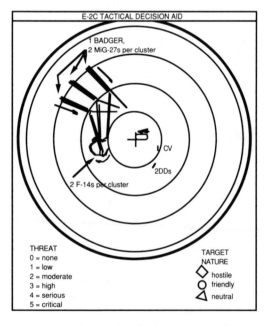

Figure 3. Overall scenario.

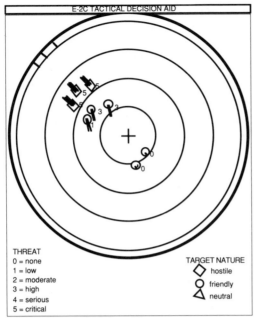

Figure 5. Stage 2 - CAP aircraft assigned to hostile targets.

the radar coverage region of the E-2C and are shown headed towards the aircraft carrier and the destroyers (surface targets). CAP aircraft are shown maintaining station since, at this stage, they have not yet been assigned to intercept the hostile targets. The threat evaluation algorithm evaluates a *high* threat to the surface targets, a *moderate* threat to the CAP aircraft,

and only a *low* threat to the hostile targets. (Note that the threat posed against the hostile aircraft is a measure of the counter-threat imposed by the friendly targets.)

The second stage of the scenario is shown in Figure. 5. The CAP aircraft have been assigned and are headed toward the hostile targets. The threat evaluation algorithm reduces the threat imposed on the surface targets to *none* based on the counter-threat provided by the CAP aircraft. Threat to the CAP aircraft themselves and the hostile targets have both been increased (as shown in the figure) based on the predicted air engagement.

Figure 6 shows the third stage of the scenario, which follows the first set of air engagements. All but two of the CAP aircraft and one of the hostile aircraft are assumed destroyed. The threat evaluation algorithm recognizes that one of the hostile targets has avoided the counter-threat provided by the CAP aircraft. This sets the threat posed to the surface targets to an increased value. In the final stage of the scenario, the surviving CAP aircraft have been reassigned and have managed to get on the tail of the surviving hostile aircraft. Accordingly, the threat evaluation algorithm decreases the threat to the surface targets. However, as in stage 2 of the scenario, the threat against the CAP aircraft and the hostile targets are

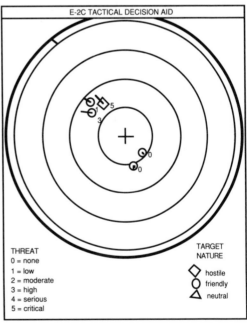

Figure 6. Stage 3 - One hostile target survives engagement.

Figure 7. Stage 4 - CAP aircraft reassigned to surviving hostile target.

increased, based on the predicted second engagement. These threat levels are indicated in Figure 7.

It can be seen that the threat evaluations made by TEM are consistent with those that are actually simulated. For the simple scenario that has been simulated, it may have been possible for an E-2C operator to predict the same set of threat conditions. However, for a more complex scenario involving hundreds of targets, it would be difficult for an operator to predict all potential engagements and the associated threats. In such situations, the threat evaluation algorithm will prove to be a significant aid to the E-2C operator.

6. Summary

We have defined a processing architecture for an airborne surveillance system that simultaneously addresses the problems of multitarget tracking and threat evaluation. We have shown that the different nature of the two problems can be addressed by a processing architecture which supports both numerical and symbolic processing. Numerical processing algorithms are based on the mathematics of estimation theory and are applicable to the problem of multisensor multitarget correlation and tracking, and identification. Symbolic processing algorithms are based on the domain knowledge of the E-2C surveillance mission and are applicable to the problem of threat evaluation.

A description of both the numerical and the symbolic algorithms has been provided. The numerical algorithms have been implemented and demonstrated in FORTRAN. Simulation results obtained with a prototype of the threat evaluation algorithm implemented in PROLOG have also been provided. These results have shown that threat evaluations made by the threat evaluation algorithm are consistent with those actually simulated.

In our ongoing effort, we plan to incorporate a more realistic database for the sensors on board the E-2C surveillance aircraft and the targets which it has to track. We also plan to integrate both the numerical and symbolic processing algorithms in the prototype of the airborne surveillance decision aid. Finally, we plan to evaluate the computational requirements of the decision aid and indicate how it might be incorporated into an operational airborne surveillance system.

References

1. "Data Fusion Research Survey Report," 1988 Data Fusion Symposium, Johns Hopkins University Applied Physics Laboratory, Laurel, Maryland, May 17-19, 1988.

2. Blackman, S. S., *Multiple Target Tracking with Radar Applications,* Artech House, 1986.

3. Bar-Shalom, Y., "Tracking Methods in a Multiobject Environment," *IEEE Transactions on Automatic Control,* vol. AC-23, August 1978, pp. 618-626.

4. Reid, D. B., "An Algorithm for Tracking Multiple Targets," *IEEE Transactions on Automatic Control,* vol. AC-24, December 1979, pp. 843-854.

5. Kurien, T., and R. B. Washburn, "Multiobject Tracking Using Passive Sensors," *Proceedings of the 1985 American Control Conference,* Boston, Massachusetts, June 19-21, 1985, pp. 1032-1038.

6. Liggins, M. E., M.A. Gerber, S.W. Gully, T. Kurien, R. B. Washburn, and M.A. Weiner, "Multispectral Multisensor Fusion Development for Enhanced Target Detection and Tracking," 1988 Tri-Service Data Fusion Symposium, Johns Hopkins University Applied Physics Laboratory, Laurel, Maryland, May 17-19, 1988.

7. Stephanou, H. E., and A. P. Sage, "Perspectives on Imperfect Information Processing," *IEEE Transactions on Systems, Man, and Cybernetics,* vol. SMC-17, October 1987, pp. 780-798.

8. Caromicoli, A., and T. Kurien, "Multisensor Target Identification in Airborne Surveillance," *Proceedings of the SPIE's 1989 Symposia in Aerospace Sensing,* Orlando, Florida (to appear).

9. Pearl, J., "On Evidential Reasoning in a Hierarchy of Hypotheses," *Artificial Intelligence* 28, 1986, pp. 9-15.

10. Caromicoli, A., and T. Kurien, "Situation Assessment in Airborne Surveillance," ALPHATECH Technical Memorandum TM-291, ALPHATECH Inc., Burlington, Massachusetts, May 1988.

15

Organizing EW Information to Support Naval Tactics

John T. Egan and Donald S. Lindberg

Introduction

The ability to acquire and apply information concerning enemy combat forces is a central problem of command and control. Because tactical EW resources collectively comprise a primary means of threat observation, this report examines the problem of EW information handling as it supports the employment of tactical actions in combat. The problem consists of how the EW information should be handled to reflect the dynamics of the operational world in a manner that is relevant for planning, for recognizing the conditions for tactical commitment, and for understanding the conditions that permit control of tactical execution. The focus of this study is to address key issues so that technical solutions may evolve.

One way of explaining the problem is in terms of the *flow, organization* and *relevance* of the information. Consider a simple model of a general force system (Figure 1). At each echelon, the tactician is required to deal with the flow of two fundamentally different kinds of information: information flowing from above and from within. The former are reports from sources outside the force, and the latter are reports from sources controlled by the force. The former tend to be more strategic or technical and broad-ranging. The latter are more detailed and voluminous. Normally the principal tactician does not control the sources from above, but does exercise full control over those from within because the information comes from organic resources.

As one proceeds downward along the battle time line, information must be organized to confront more definitive threat possibilities, leading ultimately to engagement. This means that at the lower echelons the tactician must be given tailored products relevant to his needs and of appropriate quality (accuracy, timeli-

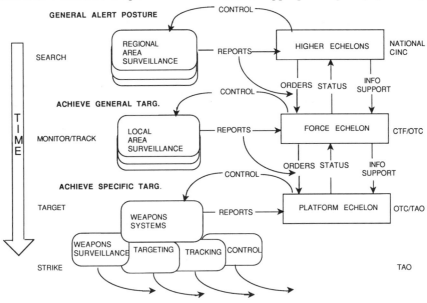

Figure 1. General force systems.

ness, coverage). How well the structure handles the flow and organization of its information and determines relevance is a measure of its effectiveness. It must be able to coordinate disparate sources of information to focus upon the same objects. It must be able to manage information so that its organization reflects the nature, status, and trends of current operations. Finally, it must provide information of the right type, in the right amount, at the right time, and to the right users to determine windows of tactical opportunity and vulnerability. The result will be the selection of tactics and countermeasures optimized for success and impact against a threat in almost any tactical situation.

The Nature of EW Information

There are three basic categories of tactical information that are available to a tactician at the force level and in some cases at the platform level—intelligence, security (cryptologic), and organic. EW information is a product of these three types as well as other classes of data which form a bounded set of information needed to formulate tactical EW plans. As such, EW information contains data from above and from within. The kinds of EW tactics supported by this information are force-wide in scope and are intended to impact the thinking of the enemy commander. They include tactics to:

• gain information (directed observation);
• withhold information (emissions control operational security, etc.);
• provide misinformation (operational deception, counter targeting, etc.);
• disrupt information (counter C³, tactical ECM , etc.);
• protect information (blocking, screening, etc.);

and have been previously discussed by Layman [1]. While terminal defense and isolated EW actions are not excluded, the approach recognizes that information games against the enemy controllers and command structure are over by this time, and attention and information is shifted toward defeating their weapons systems—and survival!

These tactics share other characteristics as well: they require persistence in their execution and consistency in the operational patterns that they expose to a hostile force. This implies a

need for further information to evaluate the effectiveness of the tactic and to replan, reposition, and refocus effort where required.

The organization of EW information is shown in Figure 2. At the top of the figure, data in the form of emitter detections, direction of arrival, and modulation parameters usually furnish the first subjects for assessments. Upon initial assessment these fall into either of two categories: ELINT or COMINT. ELINT analysis will lead to the identification of platforms and facilities and COMINT, the identification of nets and commands. As the fusion of information progresses, and further technical and combat information is brought to bear, the quality of the information reaches a point where it may be possible to gauge the progress of an ongoing threat tactic. It should be emphasized, however, that the figure represents an ideal situation where all of the data is available and there is enough time for assessments. In the vast majority of cases, the information will be incomplete.

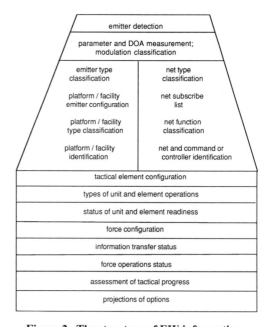

Figure 2. The structure of EW information.

EW Information Flow

The problem of EW information flow is viewed as one of obtaining the necessary information to fill in as much of the EW information structure as possible and to do so in a timely manner. This will involve not only the coordination of information from above and within but will also require interoperability between systems and analytical centers.

To be purposeful, the flow must be coordinated to support the creation of a description of the tactical EW situation. Figure 3 depicts the situation necessary for the capture of EW information that will support the analysis of tactics. The combat information arrives at the force level where it is distributed to the appropriate analytical centers. There are three basic types of analytical centers: those that provide assessments on geo-locational deployments (physical); those that provide assessments on the operational activity; and those that provide analysis of the structure and organization of the evolving threat. The products provided by these centers fall under these same categories and form what are termed *tactical indicators*.

The analysis tasks performed by these centers consist of:

• associating combat information reports to each other and/or entities of record, or new entities (an entity may be a unit, group, system, net, or combinations thereof, etc.);

• deriving assessments of current location and tracks, and detecting anomalies;

• deriving assessments of the current composition and organization of tactical units, as well as processing all data related to tactical command and multi-element system organization and composing and detecting anomalies;

• deriving assessments of the activity of the current operations of tactical units, commands, nets and weapons systems, and determining the functions performed and the status and objectives of operations, and detecting anomalies.

The products are stored in files that are distributed in a variety of locations and are referred to in the diagram as distributed analysis files.

If EW tactics goals are to be achieved, all of the information required for a tactical problem must be identified and brought into a common focus for analysis and decision making. For this reason, we establish the requirement for a new system to fuse EW information (an EW data fusion system).

The basic purposes of this system are:

• to ensure the tasking, flow and receipt of information required for analysis;

• to create a description of the tactical EW situation;

• to track the flow of EW events in that situation;

• to correlate that flow of events with a pattern of operations in the known enemy doctrine so as to understand the nature of the tactics, the status of progress, and the objective of the tactics;

• to associate this pattern with a countermeasures plan to oppose the hostile objective and to control the flow of events of that plan.

The Organization of EW Information

The EW data fusion system must provide a means to task the three types of analytical centers for filtered information about activities, changes in the tactical situation (events), conditions requiring alerts, and other specified information in their analysis files. The information that flows from these centers, as well as from other sources, would enter a common database—an *indicators database*. This would be part of an automatic information classification system. This system would provide a file structure for organizations, characteristics, tracks, and activities data (i.e., tactical indicators, on all relevant entities within the commander's area of interest. These could be organized into groups according to functions that are being performed. Such a set of files should be defined for each tactical element present or of concern. Each element present or active could be treated as the subject of an *analytical case*. In short, the indicators database and its file system should provide the means for continuing to build information concerning single entities and groups of entities as suggested by Figure 3. It is important that the majority of the arriving reports be quickly associated with their correct group. This will enable new reports to immediately effect the analysis at the highest levels.

A case is a concept borrowed from the law enforcement community, used to gather evidence about a component of the threat so that various aspects of its deployment, operations, and structure can be analyzed. A threat element is defined as a set or a collection of entities that comprise an important piece of the entire threat. A case could be active for each detected element, as well as opened for the anticipated detection of new ones. For exam-

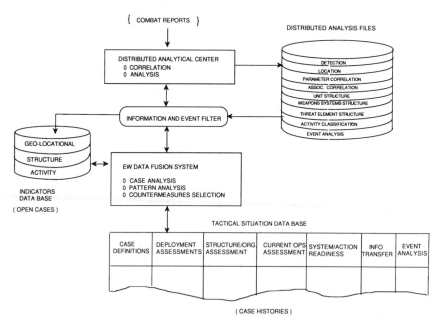

Figure 3. Capturing EW information flow.

ple, a case could be used initially to evaluate and classify data concerning the detection of platforms new to the scene. As soon as the detection is made, the analytical objective could be to determine the tactical organization to which the element belongs, the type of activity in progress, and so on. As time progresses, intermittent assessments are made about various properties of the element including its inter- and intra-actions. The sources of the information—the analytical centers, agencies, sensors, etc.—would be kept advised of the problem so that they could be tasked to search for decisive information as it is needed. The analyst, however, is not limited to attaining information from these sources only, but may also task observers and sensors directly for primary data as conditions warrant.

From the indicators database, which contains the results of first level inferences, second level inferences are made through case analyses, and from these, higher level inferences may be drawn regarding tactics. Convenient and efficient file structures must be designed and a universal vocabulary consisting of brevity codes must be invented, both for the purpose of compacting complex information in these files, thereby facilitating query and aggregation, as well as for simplifying the reporting of diverse information.

Two reasons are advanced for recommending

that the primary EW analytic data be organized in this way. The first relates to the basic differences among the three categories of data that would flow into the tactical indicators database. Each of the categories is generated by a fundamentally different observation perspective regarding what is happening. There is essentially no correlation between them of the parameters of related observable data about the underlying tactical phenomena. They are independent of each other, and analysis of one category will tend to corroborate or invalidate the results of analysis from the other. For example, intercept of the modulation parameters of a missile control signal could be an alert of a missile firing. Track data of platforms in the area can independently confirm if a launch is reasonable in terms of platforms in the area and if a missile is physically present. Activity data will reveal independently the possibility that the essential pre-launch coordination has occurred. If activity data, structural data, and track data reasonably relate, the presumption of the analysis is accepted.

The second reason for organizing input data this way is that analysis must be purposeful (i.e., case-oriented). EW tactics must be committed according to a plan of action against a finite entity; hip-shooting will simply draw the enemy and the isolated actions that ensue will provide ineffective results. The ultimate purpose of the EW data fusion system is to perform the analysis that will allow the plan to be selected and put into effect. This is the principal goal of all analysis, for which all prior analytical tasks must have laid the ground work.

In a sense, each case represents a piece of a "tactical snapshot." The collection of cases and their most current assessments will provide a more complete snapshot of the tactical situation. Of course, the total set of assessed information (from all cases) would be analogous to a set of freeze-frames, each representing a snapshot of the tactical situation. If it were possible to display these on a screen, we might see patterns of tactical activity running through them. The repository for this information is found in the tactical *situation database.*

In other words, as this information accumulates in the tactical situation database, it will form time-history patterns exposing the nature of change of tactical functions and indications of achievement. The nature of this change and the tactics being pursued provide a context for assessing the significance of the present status and projecting the probability of forms of activity. The problem, however, is that these patterns are implicitly represented in the stored information. A means must be found for extracting this information and converting it to other forms where its relevance can be determined.

Tactical Concepts

The effective organization of the information is dependent upon a proper perspective of the threat. The operations of the threat may be viewed as a continuum along the battle time line shown in the first figure. It deploys, organizes, surveys, signals, postures, targets, and strikes. All of these activities are connected and would form a continuous pattern if they could be seen. On the other hand, tactical EW information is fragmentary. It deals with indicators—bits and pieces of evidence. Different observers see different parts of the component activities.

In order to maintain this perspective, it is contended that there are only a few basic tactics components from which all combat tactics are made. They are:

- searching and analyzing
- monitoring and tracking

- targeting
- strike

Upon analysis of these generic components, it is found that they apply equally to offensive and defensive postures. In addition, there are a small number of major decisions that must be made as the battle progresses along the time line. They are: (1) decisions to be alert and maintain a monitoring posture; (2) decisions to accept the possibility of combat and to prepare for targeting; and (3) decisions to strike. The major thrust of EW information and collection should be to maintain evidence to support any of these decisions, or determining if they have been made by the enemy.

The dynamics of the situation are captured in a simple model shown in Figure 4. The model tells us that as the threat progresses along the battle time line, the enemy will provide us with the opportunity to observe and understand his activities and that we can place his individual actions within the context of an historical pattern. If our own tactical information is arranged properly, we should be able to follow his progression step-by-step, predict his next moves and apply appropriate countermeasures to delay or thwart his actions. The model also tells us that unless the hostile tactician reacts irrationally and, for example, jumps directly from search into strike (as he might do if complete surprise were desired), then he will transition sequentially and predictably until he either engages or the engagement is broken off.

What then is the nature of the pattern that will allow us to follow enemy activities and determine when it is proper to apply a countermeasure? Figure 5 shows a generic pattern and

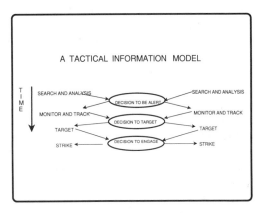

Figure 4.

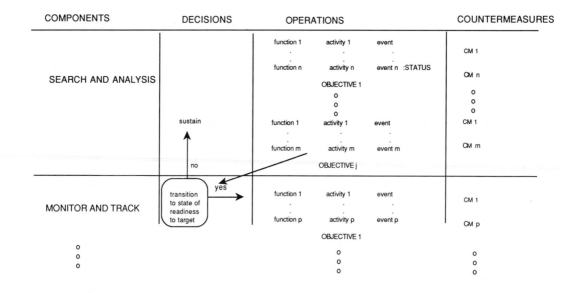

Figure 5. Tactical momentum and force level patterns.

a model of how it might work. It states that a combat operation progresses by successfully achieving a series of objectives. In the surveillance phase, for example, the objective might be to find the general location of the battle group. Once this is done, the next objective might be to identify major combatants, and so on. Each objective is achieved by the successful completion of individual steps or *force-level functions* that lead to the achievement of the objective.

Each function generates an *activity* which is observable. The successful completion of a function is termed an *event*. An event is a measured change in the tactical situation, which could be signified by the halting of a current activity and/or the beginning of a new one. The functions affiliated with an objective form a *group*. It is possible to affiliate a *status indicator* with each group, which informs us how far along the enemy has progressed toward the achievement of the group objective. The figure also shows that each function in the pattern also has an appropriate countermeasure.

In addition to the serial aspect, combat operations also exhibit cyclical patterns. This occurs when a decision has been made not to progress into the next phase of operations, but

either to sustain the current phase or to revert back to a previous one.

Determining Information Relevance

The patterns that we have described in Figure 5 represent hostile operations *explicitly*. Unfortunately, the information embedded in the tactical situation database is not compatible with this form of representation. Fusion techniques are needed to extract the implicit patterns and to process this information in a way that will allow us to understand things well enough so that countermeasures can be applied against the entire force. The success of these countermeasures will provide the ultimate test of information relevance. The understanding of this information is equivalent to performing a threat analysis. That is, a complete characterization of the threat in a way that will allow us to determine the threat-level of the current tactical situation and to project its direction.

Threat Analysis

There are several different patterns depending on the level of conflict. During a confrontation phase, of special interest are the few main patterns of functional activity, organization and compositions, deployment and stationing, that expose the combat situation (i.e., the status of attainment of the principal combat related objectives). In a general situation involving forces and commands, we submit that there may be as many as six basic patterns.

Patterns of surveillance and targeting readiness: Organization of the recent history of evidence on surveillance and monitoring activity that enable that activity to be assessed, and in particular to detect whether the enemy has attained a state of information readiness to position for targeting and engagement when desired.

Patterns of command direction that expose the potential for engagement: Organization of the recent history of evidence on dissemination of intelligence to combat forces, and intercommand liaison among forces and between forces and headquarters, to enable the nature of that collective activity to be assessed, and in particular to detect the existence of a state of requirement to target and engage.

Patterns of deployment toward potential engagement postures: Organization of the recent history of evidence concerning the deployment of hostile units and organizations, relative to our forces, to enable deployment posture to be assessed, and in particular to detect the existence of a deployment posture meeting the requirements of targeting, or of strike for the weapons involved.

Patterns of actual on-going targeting and strike readiness: Organization of the recent evidence concerning targeting and communication of target data and targeting assignments, to enable assessment of that activity, and in particular to detect the achievement of an information state indicative of readiness to strike.

Patterns of weapons readiness: Organization of the recent evidence on intercommand liaison and the overall context of activity to enable assessment of that activity, and in particular, the estimation of the intent to release weapons.

Patterns of strike action: Organization of the recent history of evidence on weapons action and weapons-support action (ECM and other combat direct support) to enable assessment of that activity and in particular to determine the status, thrust, and targets of each strike component.

Since these patterns lie buried within the tactical situation database, their processing must be continuous and executed in near real time.

As new reports arrive, they must be quickly associated with their proper groups and brought to bear upon the analysis. This will require some new computational concepts. It will require tasks or modules, designed to recognize the various patterns operating concurrently upon the database. At any given time, some of these patterns will be non-existent (null). At other times, especially those leading to engagement, all might contain valuable information. These patterns will be interrelated and if fused properly will reflect an estimate of the current threat situation and will provide the context for which to interpret new information as it arrives.

Friendly Operations Analysis

The processing of these patterns are at the heart of threat assessment, but is by no means the complete story. We must also factor in information about our own forces, because it is action relative to our deployment that sets the context for battle. This entails consideration of the status of our own operations, assigned mission requirements, requirements for deception and security, environmental constraints, current deployment, and the rules of engagement and current priorities of engagement. This marrying of threat analysis with own operational analysis constitutes a joint analysis, resulting in a classification of the current situation (i.e., a current tactical picture with an estimate of the enemy function, objective, and the status of attainment) displayed in a convenient format. At this point, projections of threat options can be made by applying templates of situation-dependent hostile doctrine (i.e., comparing the evolving pattern against a template pattern in the known doctrine of the enemy. These templates would spell out how the enemy is likely to pursue his objectives under the circumstances. If templates are not available, then other empirical evidence gained through prior intelligence, or skilled judgement must be substituted. A similar projection will be made to determine the best friendly options to follow and which will be matched against the enemy's possibilities.

The flow diagram shown in Figure 6 illustrates this process. Under ideal circumstances, this third level of analysis is expected to produce benefits consisting of matched readiness reflected in upgraded tactical deci-

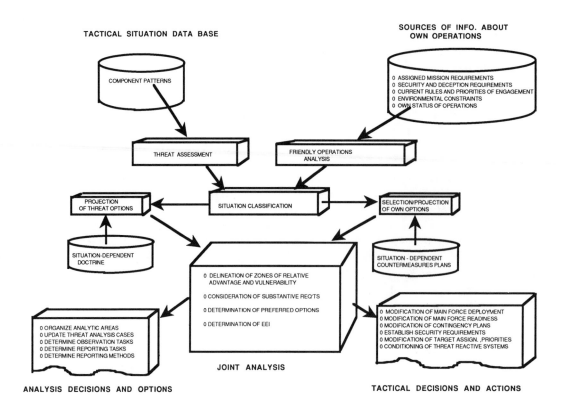

TACTICAL SITUATION DATA BASE

SOURCES OF INFO. ABOUT
OWN OPERATIONS

Figure 6. Third level tactical analysis.

sions; and upgraded analysis decisions suited to the chosen course of action.

An Example

To illustrate some of the concepts, a simple example is given. These three messages have been received:

OTC NET ACTIVE. CODED COMMS FROM OTC TO CRUISERS XY AND YZ AND DESTROYERS ABC AND DEF. CODE FORMAT 123 IS USED. TIME 1435Z.

TARGETING RADAR ACTIVE. SINGLE SWEEP. HELO BORNE AT 1410Z. POSIT 10W 48N. WITHIN TARGETING RANGE

MOVEMENT OF TWO CRUISERS AND TWO DESTROYERS TOWARD BLUE FORCE AT 1500Z.

What is actually conveyed by the reports? Do they signify a transition from surveillance into targeting? Or do they merely indicate that a targeting resource is being used to determine new locations for continued surveillance? Since the messages are out of time sequence, how do they relate with respect to the progression of tactics?

To some degree all hostile activity can be observed, but many observations are not definitive of the underlying capabilities, deployment, or objectives. In general, no single observation is decisive in terms of what is really happening. So, by itself it does not enable a coherent CM act. Observations generally create a fog of knowledge.

This fact was first noted by Clausewitz [2] who stated:

"Many intelligence reports in war are contradictory; even more are false, and most are uncertain. The tactician must possess a standard of judgement, which he can only gain from knowledge of men and affairs and common sense."

Most reports are essentially momentary views that are very limited in scope. If we are to make the best sense possible of such reports, they must be grouped as they relate to functions in progress, and the functions as they relate to overall tactics. Individual reports must be viewed in the context established by both groups.

Let us show how this can be done. Starting with the detection of the helo, we first examine the patterns stored in the tactical situation database. A pattern of surveillance and targeting readiness informs us that the blue force has been under intermittent general surveillance for the previous 48 hours. We then open a case analysis with the helo as the subject (the tactical threat element) of the case. An assumption is made that the helo is being used to target the friendly force.

The geo-locational and organization indicators support the hypothesis that the targeting of force elements is under way, because targeting capable platforms are in position and are configured to do so. However, the activity indicators do not support this premise. The OTC net is not normally used for targeting operations and coded format 123 is used primarily to transmit position and track data, not stationing or fire orders. The quality of information suggests that while enemy operations are being conducted covertly, the tempo of operations, frequency of data refresh and general operational readiness of the hostile force do not suggest that an engagement is imminent.

Conclusion: we reject the targeting hypothesis. It is assumed that the action supports an attempt to maintain contact. Case closed! But not before the various assessments are stored in the tactical situation database and become part of the recent surveillance history to show that events have progressed from determining the general location of the blue force, possibly using long range sensors, to establishing contact with the main body.

In this example, the conditions were simple enough that a complete analysis with joint assessments was not needed, although a knowledge of own posture was implied. In more complicated scenarios, involving many threat elements, engaged in what appear to be disjointed (and deceptive) operations at different levels of combat, more sophisticated analyses will be needed to put the puzzle together.

Conclusion

This report has tried to achieve three objectives. The first is to reduce an extremely complex information management problem to understandable dimensions within a systems framework. When the problem was decomposed into the flow, organizational and relevance aspects, distinct technical as well as operational challenges were exposed. It is hoped that this will be useful in organizing efforts to attack the problem in a coherent manner.

The second objective was to define the requirements for an EW data fusion system that would function at the force level in a pro active and continuous manner and would possess the generality to analyze the application of any tactical countermeasure. Force level prosecution of EW is a relatively new and still evolving Navy concept. There is currently no existing or proposed system to fuse EW information and permit the mind of the commander to focus clearly on the tactical situation.

The approach is not without pitfalls. The models described assume an ideal world—a world ruled by logic. In the real world, opponents often begin logically, but then stumble upon each other, and in the resulting melee, order and logic are replaced by chaos. Nevertheless, it is useful to try to understand the nature of combat as if it were based upon logical and orderly principles. The design of effective battle management systems could not proceed otherwise. We must ensure, by close partnership with the user, that the designed system reflects his needs and will be flexible and user-friendly enough to function under duress.

With a systems concept in place, we are ready to pursue technical solutions, knowing, of course, that there may be boundaries to what technology can ultimately provide without human partnership. In the words of Clausewitz [3]:

"......Things are perceived, of course partly by the naked eye and partly by the mind, which fills the gaps with guesswork based upon learning and experience, and thus constructs a whole out of the fragments that the eye can see; but if the whole is to be vividly present to the mind, imprinted like a picture, like a map upon the brain, without fading or blurring in detail, it can only be achieved by the mental gift that we call imagination."

If this work can be reduced to a single idea it is this: By providing the best available information in a timely manner, the imagination of the commander might be applied earlier and more effectively.

References

1. Layman, G.E., "C³CM: A Warfare Strategy," *Naval War College Review,* March-April 1985, pp. 31-42.
2. Clausewitz, *On War.* M. Howard and P. Paret (eds. and trans) Princeton University Press, (1984) p. 109.
3. Ibid., pp. 109.

16

The Potential Application of Neural Networks and Neurocomputers in C³I

Ronald E. Wright

Development of Naval Command and Control Systems

Since the 1950s, we have seen the increasing use of digital techniques to present a tactical situation picture to the ship's combat teams, but these systems give little support to the crew in assessing the situation and making appropriate decisions. The human operator is still an essential component in most command and control systems. However, we are beginning to see applications where the situation or performance requirements preclude the use of human operators-in-the-loop, for example, in the control of naval close-in weapon systems (CIWS) and battle management stations for ballistic missile defense.

The technology of the Fifth Generation of computer systems, which addressed the representation and processing of knowledge, appeared to offer a means of providing intelligent support in command and control processes, particularly by the use of expert systems [4]. Papers presented at the earlier MIT/ONR workshops [1-3] have described the use of expert systems in command tasks such as task force disposition at sea, flying programs and EW management. These systems have a rule-based approach, and enable procedures and experience to be built into a system. Thus, expert systems are really a design discipline; they do not give the system a creative intelligence. In fact, unless they are carefully designed they may give the system an unwanted rigidity and predictability. Over the last five years such systems have been tested with both the British and U.S. navies, but we are still some years away from having such systems incorporated into operational equipment.

Naval personnel and their children now look to digital computers, usually in the form of a pocket calculator, for their numeric calculations. For our symbolic expressions, the mighty pen is being replaced by the more powerful word processor. Soon we can expect to look to expert systems as planning and diagnostic aids. What we now seek is help for the operator in other creative functions, such as situation assessment and decision making. Such help appears to be at hand in the form of neural networks or neurocomputers.

What properties has the human brain that make it so essential in command and control?

• It has awareness of what is going on, partly by references to its own built-in world model.

• It has the ability to focus on the essentials of a situation, direct its attention selectively and work with incomplete or uncertain information.

• It has short and long-term memory, with an ability to associate relevant information. It is particularly good at recalling temporal and spatial patterns and sequences, such as music and road routes.

• It can exercise judgement and make decisions, responding to real-time deadlines.

• It has wishes and works toward goals—it cares!

• It is adaptive and creative.

On the negative side:

• It can be distracted or get tired and inattentive.

• It can become confused and get the wrong

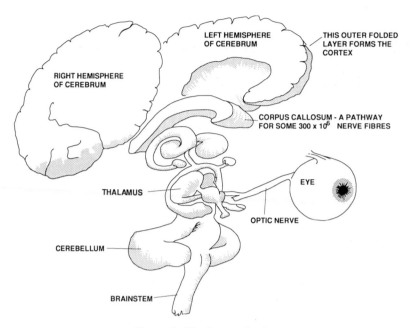

Figure 1. The human brain.

end of the stick (sometimes because of limitations of its world model).

• It can be emotional and illogical under stress.

• It is fallible.

• It can become ill, be injured and die .

• It needs a finite time to respond.

• It comes complete with a body that has to be fed, watered, and sheltered.

So, in some military systems we need to augment the human brain or replace it. Here I introduce my First Law of Neural Networks:

Human intelligence will only increase by improved human teaching techniques and the development of real creative artificial intelligence.

Figure 1 shows the human brain in partly exploded form. There is nothing else like it in the known universe. As this illustration shows, it has a highly complex architecture. It has been studied and probed by specialists from many disciplines and a vast quantity of research papers and books have been written about it. But if we ask how it works, we have only clues and not real answers. [5,6].

Some progress in research has been made at the parts level, particularly the nerve cell or neuron (Figure 2). The neuron, which is just about resolvable by the human eye, has inputs (the dendrites) that receive information and a

cell body which processes the information to give an output (in the form of an electrical pulse) to other neurons and cells along its single axon. The axon separates into a number of small fibers that have terminals or buttons. Each of these terminals forms a functional connection to another cell, called the synapse.

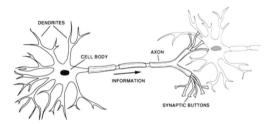

Figure 2. A typical neuron.

As far as we know, a neuron communicates with other neurons (or muscle or gland cells) only by way of these tiny synaptic junctions. A given neuron in the brain may have several or many hundreds of connections to other neurons. If the human brain has 10^{11} neurons, it might then have 10^{14} synapses and 10^{14} factorial as the number of possible connection arrangements. It appears that it is in this mas-

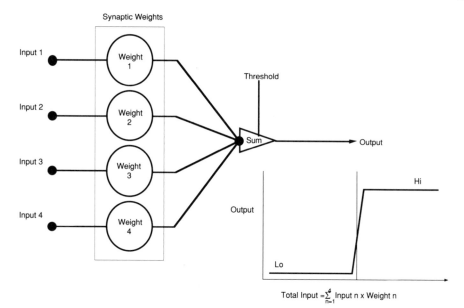

Figure 3. Single layer artificial neuron.

sive potential interconnectivity that the brain has its ability to store and process information.

One approach to understanding how the brain functions is by computer modeling at the neuron level. Figure 3 shows a simple model of a single neuron having four inputs and one output.

The inputs are connected to summing units through a device that applies a set of synaptic weights. The output depends on the magnitude of the inputs modified by the synaptic weights and the transfer characteristic of the "summer," which represents our neuron cell body or processing element. The neuron fires with its output going high when the summed inputs exceed a threshold. This neuron could be adjusted or "trained" to fire, for example, when any two or more of its inputs are in the high state. It is a simple matched filter.

An artificial network can be formed by interconnecting a number of neurons (Figure 4) and current research is directed to defining types of artificial networks and network elements and investigating how such networks can be adjusted or trained to do specific functions, such as pattern recognition. For example, the network illustrated can be adjusted to classify a four bit input pattern.

Artificial networks are not constrained to just mimic the networks and components of the human brain. A number of types of networks have been postulated and evaluated, mainly by modeling on some form of special processor or supercomputer, which at some level uses serial Von Neumann type operations. Such networks (connectionist networks) are typified by their:

- network topology
- activation rule
- learning (programming) rule
- dynamics

There is now a considerable but confusing amount of literature describing over thirty different classes of networks [7,8].

The above discussion has used the terms neural network and neurocomputer without defining them. It appears to be generally accepted that a neural network is the interconnected set of physical components such as that illustrated in Figure 4, and it could be put in a black box with a set of input and output terminals. A neurocomputer contains one or more neural networks, but also provides its user with means of specifying, controlling, running, and monitoring the neural network. A virtual neurocomputer is similar, but the neural network is modeled or emulated rather than being a direct physical representation, and the modeling may not be in real time.

Research Objectives

It appears certain that neural network technology will have an impact on future defense systems, but where they will be best applied

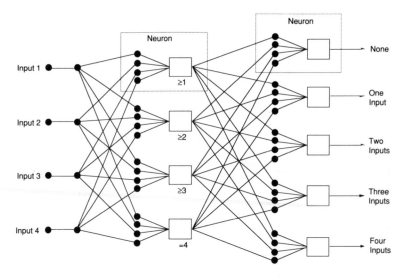

Figure 4. Two layer artificial neural network.

and how they will be realized technically is a matter of some speculation [9].

The pragmatic designer of command and control systems will want to know:

• Where neural networks can advantageously be applied

• Which type of network to use in a particular application

• How to design networks into a system and predict their performance (architectures, tools, and design methodology)

• How to implement the networks in hardware and programming terms

• How to test and qualify the resulting system

One approach would be a top-down functional decomposition of the system requirements, provided that some means could be established to identify the functions which lend themselves to neural network implementation. The difficulties of dividing the C[2] process into component parts have been addressed by Levis and Athans [10]. However, the simple conceptual model of Figure 5 will serve our present purposes. We anticipate that progress can be made in the immediate future by contacts between people conversant with C[2] design and people who, through hands-on modeling, are beginning to appreciate the different types of neural networks. What is needed is an explicit design methodology. As a contribution, I offer my Second Law of Neural Networks:

Any process that is better performed by a human or animal brain than by a computer is a potential application for neural networks.

I leave the reader to apply this law to the command and control processes.

The rest of the system designers' requirements are still in a process of evolution, and indeed some aspects and problems have not yet been defined or addressed. They thus form an area for research and development. This is not to say that practical implementations of neural networks are not already possible. In fact, it has been the success of some experimental applications of neural networks in areas where conventional AI and symbolic and numeric computing have been found wanting that has led to the current resurgence of interest in neural network techniques.

One way in which progress can be made is to focus expertise in neural networks onto one or

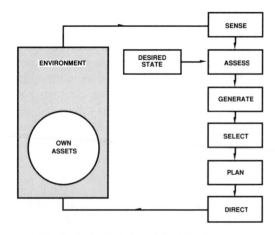

Figure 5. A simplified model of the C[2] process.

more potentially realizable applications. Fortunately, the second phase of the European Strategic Programme for Research and Development in Information Technology (ESPRIT) was launched in 1988 and a small number of projects involving the study and demonstration of the applications of neural networks have been initiated.

In 1987 the Tactical Technology Office of the U.S. Defense Advanced Research Projects Agency sponsored a comprehensive study of neural networks under the direction of the MIT Lincoln Laboratory [Ref. 11]. Its objectives were to identify potential applications for neural networks in DOD systems; to determine the neural network technology base; to identify technology requirements; to identify a DOD program plan for the following five years. The published report of the study is a comprehensive review of the state of the art at the beginning of 1988. It gives not only the enthusiastic visions of the promise of neural networks but the critical appraisal of those willing to acknowledge the limited scope of actual research efforts.

It is informative to try to determine from the DARPA study the extent to which neurocomputers have so far demonstrated properties which can begin to support or replace the essential nature of the human brain in command and control that we discussed earlier.

Several forms of adaptive behavior have been demonstrated in artificial neural networks. A fundamental feature which appears to distinguish them from any other sort of computing developed thus far is *trainability*. That is, behavior is learned from examples rather than a set of fixed instructions or rules.

Some of the more sophisticated neural network models include both *short and long-term associative memory* mechanisms for spatial and temporal patterns.

Some neural network models have been shown to find approximate solutions to certain optimization problems. However, this aspect requires further study to be of more general value in optimal decision making.

Most studies to date have focussed on the direct learning of tasks. However, there are examples in which the capability of neural networks to develop *internal models* of an external environment (world model) has been shown.

Although the importance of mechanisms to provide *selective attention* in a neurocomputer is recognized, this area has yet to be fully addressed. The ability to rapidly assess the important features of a given situation seems in many ways related to the problem of selective attention, and this area remains to be explored.

Motivation or goal-driven problem solving is an area not yet addressed by neural networks. This is presumably because it is a relatively high-order or abstract process (involving various other processes) in neural network terms.

Creativity is another area not yet addressed by neural network research. This area also seems to comprise various interacting processes. Reasoning by analogy, prediction (in the most general 'what-if' sense), and other complex processes seem to be involved in creativity.

We are a long way from replacing our human operator, but the programs of work being initiated indicate that most of the needs of the system designer are likely to be researched and developed. One downstream problem, which may not yet be receiving the attention it deserves, is how you test and validate a system incorporating neural networks. Conventional computer systems are vulnerable to failure of a single component or software bug, leading us to resort to redundant dissimilar software. Neural networks, by their massively parallel organization, are inherently tolerant to single or even multiple component failures. However, they are usually programmed by training procedures, so that different neurocomputers trained for the same task may not have identical structure and parameter values. Even if the design is performed on a virtual neurocomputer and the design is manufactured by replicating identical neural networks and parameter values, it is often by no means apparent how the network is organized and what algorithms it is using. If we give our quality controller a listing of, say, ten million interconnections, what is he going to make of it?

We ourselves have some insight into this problem, as people are a form of neurocomputer and we have been training them and assessing their capabilities for years. So our neurocomputer can be trained, tested, and gradually promoted to having increased responsibility in a similar way to human operators.

So, I introduce my Third Law of Neural

Networks, namely:

A neural network/neurocomputer is entitled to the same consideration and promotion as its human counterpart.

Conclusions

Neural network technology appears to be complementary to numerical and symbolic computing and offers the potential of helping or even replacing the more creative activities of human operators in C³I systems. Similar expectations were held for knowledge based systems, but these have not been fully realized. Artificial neural networks, because of their analogy to the neurons of the human brain, must have the potential for ultimately producing artificial creative intelligence. The immediate future offers the possibility of mechanizing applications which can be based either on a low level of neural network implementation or networks of such a size as can be represented on available virtual neurocomputers. However, our current dominant computer technology is based on two-dimensional planar semi-conductor technology, which does not lend itself to massive interconnectivity. Thus large fully-implemented neural networks await major developments in component technology.

The brain is highly structured at various levels, and these higher level architectures are not fully understood. Most models of neural networks assume a uniform homogeneous structure, and the significance of heterogeneous networks and higher level architectures has yet to be determined. Such considerations indicate that the development of neural networks to extend the areas of mechanization of C³I will be a protracted and difficult task. However, neural networks may eventually represent the high ground of command and control technology.

The Japanese, with their study of automated brain functions in their ambitious "Human Frontier Science Programme," have quietly introduced the Sixth Generation of computing.

So, I introduce my Fourth Law:

There will be no further computer generations after the Sixth, just extensions of it.

We have reached a final frontier in man's computing endeavors. In a few years time your battle commander will reach for his neurocomputer, or more likely speak to it, whenever he has a tactical decision to make. Moreover, as our understanding of the mechanisms of the mind increases, it will become apparent that we ourselves are just robots, complex and capable as we are. Hopefully by then the idea of resolving conflicts by engaging in battles of robots against robots will have become a thing of the past, and perhaps conflicts will be resolved by some omnipotent authority, possibly in the form of a multinational neurocomputer called Solomon.

Acknowledgments

I would like to acknowledge the help of Ken Lees in preparing this paper and the enthusiastic support given by Basil de Ferranti to many U.K. researchers in this field until his untimely death in 1988.

References

1. Wright, R. E. "An Overview of How Expert Systems May Be Applied in Naval Command and Control Systems." *Proceedings of the 8th MIT/ONR Workshop on C³ Systems*, 1985. MIT Reference LIDS-R-1519, pp. 7-13.

2. Gadsden, J. A. and Lakin, W. L. "FLYPAST : An Intelligent System for Naval Resource Allocation." *Proceedings of the 9th MIT/ONR Workshop on C³ Systems*, 1986. MIT Reference LIDS-R-1624, pp. 139-144.

3. Lakin, W. L. and Miles, J. A. "Intelligent Data Fusion and Situation Assessment." *Proceedings of the 9th MIT/ONR Workshop on C³ Systems*, 1986. MIT Reference LIDS-R-1624, pp. 133-138.

4. Taylor, Edward C. "AI in Command and Control: What and When." *Proceedings of JDL Command and Control Research Symposium*, 1987. Reference SAIC-87/1119, pp. 379-384.

5. Ornstein, Robert and Thompson Richard F.,*The Amazing Brain*, London: Chatto & Windus, 1985. ISBN 0-7011-3989-8.

6. Searle, John. Mind, *Brains and Science*, London: British Broadcasting Corporation, 1984. ISBN 0-563-20222-1.

7. IEE First International Conference on Neural Networks, San Diego: SOS Printing, June, 1987. Four Volumes. IEE Cat. No. 87THO191-7.

8. Eckmiller, Rolf and Malsburg, Christoph v. d., (ed.) *Neural Computers*, New York: Springer-Verlag, published in cooperation with the NATO Scientific Affairs Division, 1988. ISBN 0-387-18724-3. NATO ASI Series, Series F, Vol. 41.

9. North, Robert L. "Neurocomputing: Its Impact on the Future of Defence Systems," *Defense Computing*, Jan.-Feb. 1988, pp. 18-23, 66.

10. Levis, Alexander H. and Athans, Michael. "The Quest for a C³ Theory: Dreams and Realities." *Proceedings of JDL Command and Control Researc h Symposium*, 1987, pp. 7-11. Reference SAIC-87/1119.

11. *DARPA Neural Network Study*, Fairfax, Virginia: AFCEA International Press, 1988. ISBN 0-916159-17-5.

17

Mathematical Comparison of JANUS(T) Simulation to National Training Center Data

Lester Ingber

Introduction

Necessity of Comparing Computer Models to Exercise Data

This project addresses the utility of establishing an approach to compare exercise data to large-scale computer models whose relatively underlying microscopic interactions among men and machines are driven by the natural laws of physics.

In this paper, the term computer model will be used to include computer simulations as well as computer war games, the latter involving human participants in real time. (It appears that JANUS(L) simulation can be compared favorably to JANUS(L) war game). In this study, the focus will be to compare JANUS(T) war game to National Training Center (NTC) data, since both systems then take into account human interactions.

It also should be noted that "large-scale" here refers to battalion level. (Army systems scale by factors of 3-5, from company to battalion to brigade to division to corps to army.) If these battalion level computer models can be favorably compared, and if consistency can be achieved between the hierarchy of large-scale (battalion level), larger-scale (corps level), and largest-scale (theater level) computer models, then these higher echelon computer models also can be favorably compared. This could only enhance the value of training on these higher echelon computer models.[1]

The necessity of depending more and more on combat computer models (including simulations and war games) has been brought into sharper focus because of many circumstances, for example: (a) the nonexistence of ample data from previous wars and training operations, (b) the rapidly shortening time scale on which tactical decisions must be made, (c) the rapidly increasing scale at which men and machines are to be deployed, (d) the increasing awareness of new scenarios that are fundamentally different from historical experiences, and (e) the rapidly increasing expense of conducting training exercises.

Furthermore, such computer models could be used to augment training. We spend several million dollars to cycle each battalion through NTC. The training of these commanders could be greatly enhanced if relatively inexpensive pre- and post-training war games were provided that could statistically replicate their training missions. Even, or rather especially, for the development of such training aids, proper analysis and modeling is required to quantitatively demonstrate that the computer models are good statistical representations of the training mission.

However, the level of acceptance of computer models in major military battle management and procurement decisions appears to be similar to the level of acceptance of computer simulations in physics in the 1960s. Prior to the 1960s, theory and experiment formed a close bond to serve to understand nature. In the 1960s, academicians were fascinated with evolving computer technology, but very few people seriously accepted results from computer simulations as being on a par with good theory and good experiment. Now, of course, the situation is quite different. The necessity of understanding truly complex systems has placed computer simulation, together with theory and experiment, as an equal leg of a tripod of techniques used to investigate physical nature.

The requirements necessary to bring combat computer models to their needed level of importance are fairly obvious. In order to have confidence in computer model data, responsible decision makers must be convinced that computer models model reality, not metaphors of reality; models of models; or models of models of models, etc. Many people feel that not much progress has been made in the last decade[2,3] with regard to this issue, despite a general awareness of the problem.

If a reasonable confidence level in computer models of large-scale combat scenarios could be obtained, there are several immediate payoffs to be gained. More objective data could be presented for procurement decisions, e.g., provided by sensitivity analyses of sets of computer models differing in specific weapons characteristics. In order to give proper weight to these differing characteristics, their influence within the global context of full combat scenarios would be tested.

Large-Scale C[2] and Need for Mathematical Modeling

Modeling phenomena is as much a cornerstone of 20th century science as is collection of empirical data.[4] In essentially all fields of science, mathematical models of the real world become tested by fitting some parameters to empirical data. Since the real world is often nonlinear and stochastic, it is not surprising that often this fitting process must involve fitting statistical, nonlinear, functional forms to data.

As in other fields of science, in the context of modeling combat, reductionist doctrine is simply inadequate to fully understand large-scale systems. For example, a threshold is quickly reached at a level of any large system, be it physical, biological or social, when a "language" shift is required for effective command and control. A high level commander cannot use a grease board to track individual units, albeit he might periodically sample his units; rather, he must look at the overall systematics, that is, aggregated measures of force (MOF) or effectiveness (MOE), attrition, resupply, etc. At this level we properly require command and control (C[2]), rather than "supra-battle management" from commanders. At this level we denote the system as large-scale. (See Figure 1.)

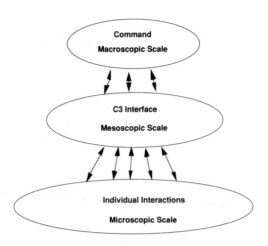

Figure 1. Scaling C[3] systems.

This issue of utilizing MOFs and MOEs (e.g., starting at about battalion level of combat) is relevant to computer models as well as to actual combat. Merely aggregating data to form MOFs or MOEs does not determine if results from one mission (combat or computer model scenario) are comparable to another mission. For example, small differences in tempo or in spatial distribution of FLOT (forward line of own troops) or FEBA (forward edge of the battle area) may cause tables of numbers to appear quite different.

Mathematical models of aggregated data should be expected to uncover "mechanisms" of combat, such as line-firing or area-firing, in simple Lanchester theory. More complex missions plausibly will contain more subtle mechanisms as well as weighted contributions of more basic mechanisms. Using this as hindsight, in some systems it may then be possible to specify a figure of merit; some simple set of numbers to capsulate the influence of these mechanisms.

These mechanisms are to be articulated by particular algebraic forms. Indeed, this is the most important sense of physics, to use mathematical forms to articulate mechanisms and models of empirical phenomena. For example, it is to be expected that the use of this statistical mechanical approach will facilitate the process of identifying algebraic forms relevant to combat, because in many cases these algebraic forms will be sufficiently similar to other well-known physical mechanics determined in other physical studies.

Awareness of such plausible mechanisms

permits the analyst or the field commander to more easily uncover variations of patterns or common themes unfolding in databases or in real time. In complex missions, such tools are more than niceties; they can be necessities.

Therefore, to be most useful, computer model data should be aggregated and then mathematically modeled using variables as close as possible to the level of command to which they are to become decision aids. The mathematical modeling must by necessity be nonlinear, that is, offering alternative choices of possible outcomes, upon which human judgment and experience can be brought to bear and be accountable. The mathematical modeling must by necessity also be probabilistic, so that expected gains can be humanly weighed with respect to both payoffs and the probabilities of payoffs of alternative dynamic (time-dependent) states of the system.

It must be emphasized that this approach requires an evolution of knowledge. This project is developing models suitable to describe the statistical nature of selected force-on-force battalion and brigade scenarios. It is expected that the accumulation of models of many types of scenarios will lead to a better fundamental understanding of combat with direct operational applications.

As is often discussed,[5] too often we have weighted the communications aspect of C^3, to the detriment of not properly addressing the command and control aspects, that is, too often only seeking technological fixes to hard large-scale problems. For example, the Soviets give much weight to C^2, and they have structured tabular decision aids dispersed through their levels of command. However, their relatively rigid political mind-set has fostered the development of these tables by using quasi-linear, essentially deterministic mathematical models fitted to operations data. If we use modern methods of nonlinear stochastic mathematical modeling, using data gained from operations as well as from more advanced computer models baselined to actual operations, then we can greatly increase our C^2 advantage.

Need for Mathematical Models of Aggregated Combat Data

The reasons for seeking mathematical models of exercise data correspond to the same reasons for seeking mathematical models of computer models, as described in the previous section.

These mathematical models can be used to approach comparison of computer models (e.g., JANUS[T]) to exercise data (e.g., from NTC). Similar mathematical expressions must describe to similar mechanisms, whether they exist in the computer models or in the exercise data. Only if computer models such as JANUS(T) can be favorably compared to NTC data, can these war games provide reliable pre- and post-exercise training for NTC commanders.

Such models also need to be developed at battalion level, to drive corps and theater level models which will require highly aggregated models to run in real time. For example, we expect good models of battalion level combat to be nonlinear and stochastic. However, such models are too mathematically complex to drive higher echelon models, especially in real time for war gaming. After this mathematical model is developed, only then is it correct and reasonable to determine decision rules, e.g., describing bifurcation of trends of the fitted mathematical distributions, and to linearize distributions in "most likely" regions. In this manner, the most salient features of properly fitted battalion level models can drive higher echelon models.

Now, the task in this approach is to find the best mathematical model of each system, that is, the computer model system and the exercise data system. Then, the best mathematical models for each can be compared.

Models Versus Reality

It must be stated that there are still many problems faced by all computer models of combat, which must be solved before they can be accepted as models of reality. For example, a very basic problem exists in the quality of acquisition algorithms, i.e., how to construct an algorithm that realistically portrays human attention (pre-attentive as well as selective) and perception, under various combat and weather conditions, night versus day, etc. The influence of attention and perception on complex physical[6] and mental tasks[7] has received considerable attention by the author. Currently, the best combat computer models treat acquisition as serial and logical processes, whereas the human brain acquires data by parallel and associative processes. Therefore, the inclusion of human players in multiple runs of similar scenarios is essential, if a probabilistic mathematical model is to be developed to model exercise data such as that obtained from NTC.

Presently, line of sight (LOS) algorithms seem to be the most costly time factor in running JANUS(T) computer models. Even if more realistic acquisition algorithms are developed, they must be tailored to the needs of real-time computer models if they are to be used in war gaming and in training.

Similarly, in order to develop a computer model of NTC exercises, to perform the comparison approach of this project, an acquisition model must be supplemented by existing algorithms within the computer model. This is a very difficult problem, requiring subjective, albeit expert, judgment.

Individual Performance

It is clear that individual performance is extremely important in combat,[8] ranging from battle management of the commander, to battle leadership of sub-commanders, to the degree of participation of individual units, to the more subtle degradation of units performing critical tasks.

Our analyses of NTC data concludes that data collected to date is not sufficient to accurately statistically judge individual performance across these scales. However, we do believe that this data is sufficient to analyze battle management, perhaps battle leadership in some cases. This is essential if we are to statistically compare JANUS(T) to NTC, and thereby prepare a what-if capability for JANUS(T) to augment NTC training.

It is important to recognize and emphasize the necessity of improving data collection at NTC, to permit complementary analyses of human factors at finer scales than our statistical approach permits.

Technical Background: General

Problems in Lanchester Theory

Quasi-linear deterministic mathematical modeling is not only a popular theoretical occupation, but many war games, such as JTLS (Joint Theater Level Simulation), use such equations as the primary algorithm to drive the interactions between opposing forces. In its simplest form, this kind of mathematical modeling is known as Lanchester theory:

$$\dot{r} = dr/dt = x_r\,b + y_r\,r\,b$$
$$\dot{b} = db/dt = x_b\,r + y_b\,b\,r \tag{1}$$

where r and b represent Red and Blue variables, and the x's and y's are parameters which somehow should be fit to actual data.

It is notoriously difficult, if not impossible, to use Equation 1 to mathematically model any real data with any reasonable degree of precision. These equations perhaps are useful to discuss some gross systematics, but it is hard to believe that, for example, a billion dollar procurement decision would hinge on mathematical models dependent on that equation.

Some investigators have gone further, and amassed historical data to claim that there is absolutely no foundation for believing that Equation 1 has anything to do with reality.[9] However, although there is some truth to these criticisms, the above conclusions do not sit comfortably with other vast stores of human experience. Indeed, this controversy is just one example that supports the necessity of having human intervention in the best of C^2 plans, no matter how (seemingly) sophisticated analysis supports conclusions contrary to human judgment.[8] When dealing with a dynamic complex system, intuition and analysis must join together to forge acceptable solutions.

The use of historical data, to disclaim any truth to the validity of Equation 1, is in itself the use of inappropriate analysis.[10] The aggregation of solitary trajectories from many different stochastic combat scenarios does not necessarily form any kind of probability distribution upon which to make statistical judgments. Two combat scenarios, that differ in even only several variables, realistically are going to be quite different scenarios, not least because the very nature of nonlinear nonequilibrium (open) competitive systems is to have opposing sides pressed to their extreme, not to their average, capabilities. Given some "maneuvering room" to distort states of an open system, indeed these states will be distorted to new "favorable" values.

Therefore, as understood from experience in simulating physics systems, many trajectories of the "same" stochastic system must be aggregated before a sensible resolution of averages and fluctuations can be ascertained. Given two scenarios that differ in one parameter, and given a sufficient number of trajectories of each scenario, then the sensitivity to changes of a "reasonable" algebraic function to this parameter can offer some analytic input into decisions involving the use of this parameter in combat scenarios.

Empirical Data

Therefore, there are two remaining issues to be resolved. The first is to find a database of a sufficient number of trajectories of the "same" system, upon which mathematical models can be built. The second is to forge an effective approach to mathematically model this data.

The numerous battalion cycles of exercises at the NTC can provide more trajectories of similar large-scale combat scenarios than any other source. However, typical of exercises whose purpose is to train and not necessarily to provide data serving analyses, this data is quite "dirty."[11] Some problems specific to exercises would not occur in actual combat. There is a tremendous amount of several kinds of data, machine derived as well as derived from human observers in the form of "take-home packages." This is not meant to be criticism of exercises at NTC. Quite the contrary, while respecting the sensitivity of this data, objective analyses for this project require a complete understanding of these problems.

Mathematical Modeling

This brings us to the next issue. What is a "reasonable" mathematical modeling approach?

It is reasonable to at least tentatively accept the experience of many commanders, whose intuitions have developed to think in terms of Equation 1. Then, the problem seems to be that the degree of their quantitative, not qualitative, insights is insufficient to detail many combat scenarios. This then becomes the job of analysis, and explicates the purpose as well as the analytic task of mathematically modeling combat data. A good mathematical model must fit the data, and also be useful as a decision aid to the commander and decision maker. Therefore, we can approach this problem by considering Equation 1 as some kind of zeroth order approximation to reality.

In the late 1970s, mathematical physicists discovered that they could develop statistical mechanical theories from algebraic functional forms

$$\dot{r} = f_r(r, b) + \sum_i g_r^i(r, b)\, \eta_i$$

$$\dot{b} = f_b(b, r) + \sum_i g_b^i(b, r)\, \eta_i \qquad (2)$$

where the f's and g's are general nonlinear algebraic functions of the variables r and b.[14,15] The f's are referred to as the (deterministic) drifts, and the square of the g's are related to the diffusions (fluctuations). In fact, the statistical mechanics can be developed for any number of variables, not just two. The η's are sources of Gaussian-Markovian noise, often referred to as "white noise." Equation 1 is a special case of this generalized set of equations, with bilinear drift and mathematically zero noise. At this time, certainly the proper inclusion of multiplicative noise, using parameters fit to data to mathematically model general sources of noise, is preferable to improper inclusion or exclusion of any noise in combat models.

The ability to include many variables also permits a "field theory" to be developed, e.g., to have sets of (r,b) variables (and their rate equations) at many grid points, thereby permitting the exploration of spatial-temporal patterns in r and b variables. This gives the possibility of mathematically modeling the dynamic interactions across a large terrain.

Support for Present Mathematical Modeling Approach

These new methods of nonlinear statistical mechanics only recently have been applied to complex large-scale physical problems, demonstrating that empirical data can be described by the use of these algebraic functional forms. Success was gained for large-scale systems in neuroscience (e.g., [12,13]) and in nuclear physics. I have proposed that these methods be used in C3.[14,18]

Thus, now we can investigate various choices of f's and g's to see if algebraic functional forms close to the Lanchester forms can actually fit the data. In physics, this is the standard phenomenological approach to discovering the encoding knowledge and empirical data; that is, fitting algebraic functional forms that lend themselves to physical interpretation.

This gives more confidence when extrapolating to new scenarios; exactly the issue in building confidence in combat computer models.

The utility of these algebraic functional forms in Equation 2 goes beyond their being able to fit sets of data. There is an equivalent representation to Equation 2, called a "path-integral" representation, for the long-time probability distribution of the variables. This short-time probability distribution is driven by a "Lagrangian," which can be thought of as a dynamic algebraic "cost" function. The path-integral representation for the long-time distribution possesses a variational principle, which means that simple graphs of the algebraic cost function give a correct intuitive view of the most likely states of the variables, and of their variances. Like a ball bouncing about a terrain of hills and valleys, one can quickly visualize the nature of dynamically unfolding r and b states.

Especially because we are trying to mathematically model sparse and poor data, different drift and diffusion algebraic functions can give approximately the same algebraic cost function when fitting short-time probability distributions to data. The calculation of long-time distributions permits a clear choice of the best algebraic functions; that is, those which best follow the data through a predetermined epoch of battle. Thus, dynamic physical mechanisms, beyond simple "line" and "area" firing terms, can be identified. Afterward, if there are closely competitive algebraic functions, they can be more precisely assessed by calculating higher algebraic correlation functions from the probability distribution.

It must be clearly stated that, like any other theory applied to a complex system, these methods have their limitations, and they are not a panacea for all systems. For example, probability theory itself is not a complete description when applied to categories of subjective "possibilities" of information.[19,20] Other non-stochastic issues are likely appropriate for determining other types of causal relationships; such as the importance of reconnaissance to success of missions.[21] These statistical mechanical methods appear to be appropriate for comparing these stochastic large-scale combat JANUS(T) and NTC systems. The details of our studies will help to determine the correctness of this premise.

Technical Approach

Complexity of Typical Problems

There are variables, spatial dimensions, and parameters that must be processed by such calculations. Typically researchers have considered only a few variables, (one or two, in one or two dimensions) with several parameters. Or, they have considered limiting cases of huge/infinite number of variables/dimensions. These problems require breaking new ground into the nonlinear nonequilibrium stochastic realm of 10, 20, or 30 dynamic variables. This number is barely large enough to give reliable analysis/aids to decision makers, yet barely small enough to be able to process good scientific calculations. We must avoid handling too many variables, which leads quickly to data overload of machines and humans. And, we must avoid doing too simplistic modeling which is, at best, unreliable for complex systems.

Current State of Algorithms

Recently, two major computer codes have been developed, which are key tools to mathematically model combat data. The first code fits short-time probability distributions to empirical data, using a most-likelihood technique on the Lagrangian. An algorithm of very fast simulated re-annealing has been developed to fit empirical data to a theoretical cost function over a D-dimensional parameter space,[22] adapting for varying sensitivities of parameters during the fit. The annealing schedule for the "temperatures" (artificial fluctuation parameters) T_i decrease exponentially in "time" (cycle-number of iterative process) k, i.e., $T_i = T_{io} exp(-c_i k^{1/D})$.

Heuristic arguments have been developed to demonstrate that this algorithm is faster than the fast Cauchy annealing,[23] $T_i = T_0/k$, and much faster than Boltzmann annealing,[24] $T_i = T_0/lnk$.

The second code develops the long-time probability distribution from the Lagrangian fitted by the first code. A robust and accurate histogram path-integral algorithm to calculate the long-time probability distribution has been developed by Dr. Mike Wehner to handle nonlinear Lagrangians,[25] including a two-variable code for additive and multiplicative cases. We are presently working to create a code to process several variables.

Relationship to Other Combat Modeling

Connections can be made to other attrition-driven models that require very simple algebraic forms to drive their computer models in real time. After fitting battalion level data with our methodology, a combination of decision rules can satisfy requirements to determine space-time regions of maximum probability from the stochastic nonlinear fits, and local linearization of probability peaks to Lanchester-type simpler algebraic forms. This application can be made to higher echelon Army models, for instance, VIC (division to corps level), FORCEM (EAC to theater level), JTLS (theater level,), JESS (division to corps level).

These methods also can be used to implement C^3 models driven by attrition equations such as C^3EVAL being developed by the Institute for Defense Analyses for the J6-F office of JCS. This will permit connection of JANUS(T) to higher echelon C^3, and/or to degrade/jam communications between war gamers.

This Lagrangian approach to combat dynamics permits a quantitative assessment of concepts previously only loosely defined. Reduction to other math-physics modeling can be achieved after fitting realistic exercise and/or simulation data.

The best resolution presently available from NTC is at the company level. Our JANUS(T) what-if model can provide better resolution, at least statistically consistent with NTC data.

NTC Prototype Mathematical Model

The mathematical model comparison process develops separate mathematical models for both the computer simulation data and the exercise data, thereby permitting a common basis for quantitative comparison. (See Figure 2.) Performing this task requires intimate knowledge of each system as well as the mathematical tools described previously.

Description of NTC

The U.S. Army National Training Center (NTC) is located at Fort Irwin, California. There have been about a quarter million soldiers in 80 brigade rotations at NTC, at about

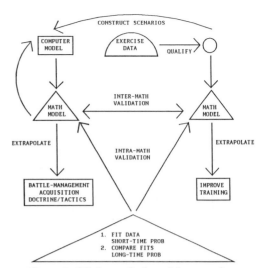

Figure 2. Mathematical model comparison process.

the level of two battalion task forces (typically about 3500 soldiers and a battalion of 15 attack helicopters), which train against two opposing force (OPFOR) battalions resident at NTC. NTC comprises about 2500 km², but the current battlefield scenarios range over about 5 km linear spread, with a maximum lethality range of about 3 km. NTC is gearing up for full brigade level exercises.

Observer-controllers (OC) are present down to about platoon level. A rotation will have three force-on-force missions and one live-fire mission. OPFOR platoon and company level personnel are trained as U.S. Army soldiers; higher commanders practice Soviet doctrine and tactics. An OPFOR force typically has ~100 BMPs and ~40 T72s.

The primary purpose of data collection during an NTC mission is to patch together an after-action review (AAR) within a few hours after completion of a mission, giving feedback to a commander who typically must lead another mission soon afterward. Data from the field, multiple integrated laser engagement system (MILES) devices, audio communications, OCs, and stationary and mobile video cameras, is sent via relay stations back to a central command center where this all can be recorded, correlated and abstracted for the AAR. Within a few weeks, a written review is sent to commanders, as part of their NTC take-home package. It currently costs about four million dollars per NTC rotation, 1 million of which goes for this computer support.

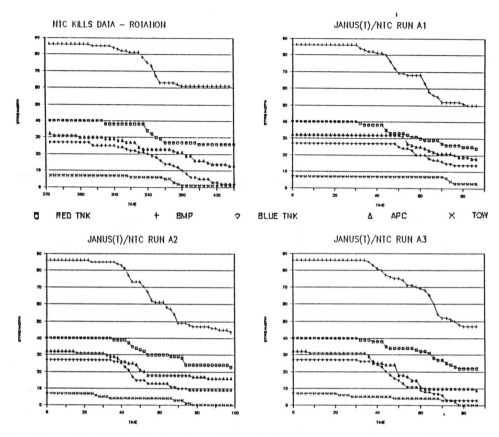

Figure 3. Attrition ("kills") data is illustrated for an NTC mission and for three JANUS(T) runs using the NTC-qualified database.

There are 460 MILES transponders available for tanks for each battle. The "B" units have transponders, but most do not have transmitters to enable complete pairings of kills-targets to be made. (New MILES devices being implemented have transmitters which code their system identification, thereby greatly increasing the number of recordings of pairings.) Thus, MILESs without transmitters cannot be tracked. Man-packs with B units enable these men to be tracked, but one man-pack can represent an aggregate of as much as five people.

B units send data to "A" stations (presently 48, though 68 can be accommodated), then collected by two "C" stations atop mountains, and sent through cables to central VAXs forming a core instrumentation system (CIS). There is a present limitation of 400 nodes in computer history for video tracking (but 500 nodes can be kept on tape). Therefore, about 200 Blue and 200 OPFOR units are tracked.

Description of JANUS(T)

JANUS(T) is an interactive, two-sided, closed, stochastic, ground combat (recently expanded to air and naval combat as an extension of our present projects) computer simulation. Interactive refers to the fact that military analysts (players and controllers) make key complex decisions during the simulation, and directly react to key actions of the simulated combat forces. Two-sided (hence the name JANUS of the Greek two-headed god) means that there are two opposing forces simultaneously being directed by two sets of players. Closed means that the disposition of the enemy force is not completely known to the friendly forces. Stochastic means that certain events, for example, the result of a weapon being fired or the impact of an artillery volley, occur according to laws of chance (random number generators and tables of probabilities of detection (PD), acquisition (PA), hit (PH), kill (PK), etc.). The principle modeling focus is on those military systems that participate in maneuver and artillery operations. In addition to conventional direct fire and artillery operations, JANUS(T) models precision guided munitions, minefield employment and breaching, heat stress casualties, suppression, etc.

Throughout the development of JANUS(T), and its Janus precursor at Lawrence Livermore National Laboratory, extensive efforts have been made to make the model "user friendly." Missing unit movements and initial force structures were filled in, often making "educated guesses" by combining information on the CIS tapes and the written portion of the take-home package. This project could not have proceeded if we had not been able to automate transfers of data between different databases and computer operating systems.

Preliminary Mathematical Modeling of NTC Data

The "kills" attrition data from NTC and our JANUS(T) simulation at once looks strikingly similar during the force-on-force part of the combat. (See Figure 3.) From the single NTC trajectory qualified to date, 7 five-minute intervals in the middle of the battle were selected. From six JANUS(T) runs, similar force-on-force time epochs were identified, for a total of 42 data points. In the following fits, r repre-

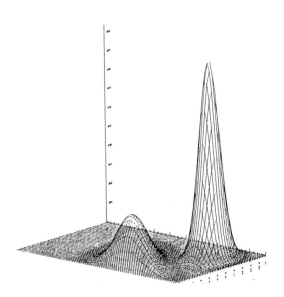

Figure 4. In this plot, the horizontal axes represent Red and Blue forces. For this JANUS(T) additive noise case, two time slices are superimposed. Taking the initial onset of the force-on-force part of the engagement as 35 minutes on the JANUS(T) clock, these peaks represent 50 and 100 minutes. Reflecting boundary conditions are taken at the other two surfaces.

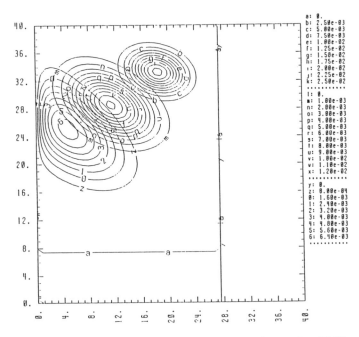

Figure 5. For the same case as in Figure 4, contour plots are superimposed at 50, 70, and 100 minutes.

sents Red tanks, and b represents Blue tanks.

Fitting NTC data to an additive noise model, a cost function of 2.08 gave:

$$\dot{r} = -2.49 \times 10^{-5}b - 4.97 \times 10^{-4}br + 0.320\eta_r$$
$$\dot{b} = -2.28 \times 10^{-3}r - 3.23 \times 10^{-4}rb + 0.303\eta_b \qquad (3)$$

Fitting NTC data to a multiplicative noise model, a cost function of 2.16 gave:

$$\dot{r} = -5.69 \times 10^{-5}b - 4.70 \times 10^{-4}br + 1.06 \times 10^{-2}(1+r)\eta_r$$
$$\dot{b} = -5.70 \times 10^{-4}r - 4.17 \times 10^{-4}rb + 1.73 \times 10^{-2}(1+b)\eta_b \qquad (4)$$

Fitting JANUS(T) data to an additive noise model, a cost function of 3.53 gave:

$$\dot{r} = -2.15 \times 10^{-5}b - 5.13 \times 10^{-4}br + 0.530\eta_r$$
$$\dot{b} = -5.65 \times 10^{-3}r - 3.98 \times 10^{-4}rb + 0.784\eta_b \qquad (5)$$

Fitting JANUS(T) data to a multiplicative noise model, a cost function of 3.42 gave:

$$\dot{r} = -2.81 \times 10^{-4}b - 5.04 \times 10^{-4}br + 1.58 \times 10^{-2}(1+r)\eta_r$$
$$\dot{b} = -3.90 \times 10^{-3}r - 5.04 \times 10^{-4}rb + 3.58 \times 10^{-2}(1+b)\eta_b \qquad (6)$$

This comparison illustrates that two different models about equally fit the short-time distribution. The multiplicative noise model shows that about a factor of 100 of the noise might be "divided out," or understood in terms of the physical log-normal mechanism.

Attrition Mean of Blue

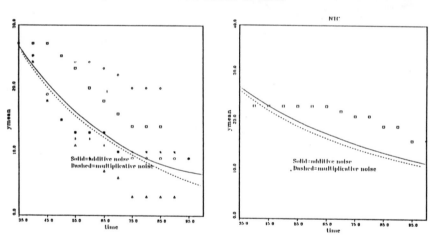

Attrition Mean of Red

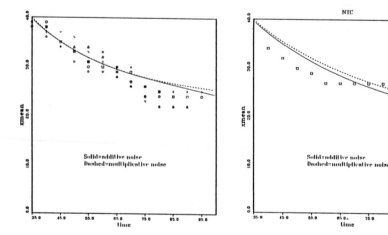

Figure 6. JANUS(T) and NTC attrition means.

Attrition Variance of Blue

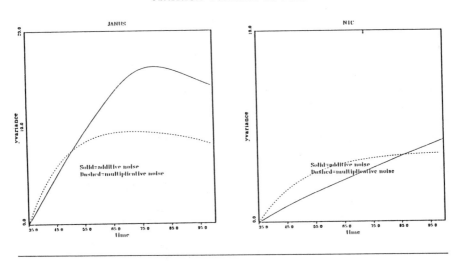

Attrition Variance of Red

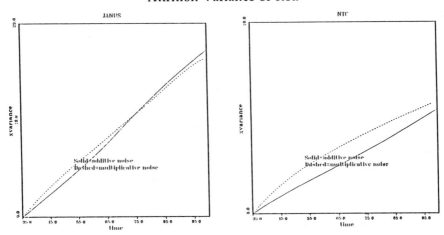

Figure 7. JANUS(T) and NTC attrition variances.

In order to discern which model best fits the data, we turn to the path-integral calculation of the long-time distribution, to see which model best follows the actual data. Figures 4 and 5 present the long-time probability of finding values of these forces. In general, the probability will be a highly nonlinear algebraic function, and there will be multiple peaks and valleys.

Figures 6 and 7 give the means and variances of tank attrition from the JANUS(T) and NTC databases. Since we presently have only one NTC mission qualified, the variance of deviation from the mean is not really meaningful; it is given only to illustrate our approach which will be applied to more NTC missions as

they are qualified and aggregated. Here, only the Blue JANUS(T) variances serve to distinguish the additive noise model as being consistent with the JANUS(T) data.

Discussion of Study

Data from 35 to 70 minutes was used for the short-time fit. The path-integral was used to calculate this fitted distribution from 35 minutes to beyond 70 minutes. This serves to compare long-time correlations in the mathematical model versus the data, and to help judge extrapolation past the data used for the short-time fits. It appears that indeed some multiplicative noise model is required. Of course, other Lanchester modelers most often do not consider noise at all, and at best just extract additive noise in the form of regression excesses. More work is required to find a better (or best?) algebraic form. The resulting form is required for input into higher echelon models. As more NTC data becomes available, we will also generate more JANUS(T) data. Then, we will be able to judge the best models with respect to how well they extrapolate across slightly different combat missions.

We have demonstrated proofs of principle, that battalion level combat exercises can be well represented by the computer simulation JANUS(T), and that modern methods of nonlinear nonequilibrium statistical mechanics can well model these systems. Since only relatively simple drifts and diffusions were required, in larger systems (e.g., at brigade and division levels) it might be possible to "absorb" other important variables (C^3, human factors, logistics, etc.) into more nonlinear mathematical forms. Otherwise, this battalion level model should be supplemented with a "tree" of branches corresponding to estimated values of these variables.

Notes

[1] G. T. Bartlett, "Battle Command Training Program." *Phalanx 21*, 18-20 (1988).

[2] Comptroller General, "Models, Data, and War: A Critique of the Foundation for Defense Analyses." Report No. PAD-80-21. U.S. General Accounting Office, Washington, D.C. (1980).

[3] Comptroller General, "DOD Simulations: Improved Assessment Procedures Would Increase the Credibility of Results." Report No. GAO/PEMD-88-3. U.S. General Accounting Office, Washington, D.C. (1987).

[4] M. Jammer, *The Philosophy of Quantum Mechanics*, Wiley & Sons, New York, NY (1974).

[5] C^3CM Joint Test Force and National Defense University, *Thinking Red in Wargaming*, Albuquerque, NM (1987).

[6] L. Ingber, *Elements of Advanced Karate*, Ohara, Burbank, CA (1985).

[7] L. Ingber, "Attention, Physics and Teaching." *J. Social Biol. Struct.* 4, 225-235 (1981).

[8] M. van Creveld, *Command in War*, Harvard University Press, Cambridge, MA (1985).

[9] T.N. Dupuy, "Can We Rely on Computer Combat Simulations. *Armed Forces Journal*," August, 58-63 (1987).

[10] W.W. Hollis, "Yes, We Can Rely on Computer Combat Simulations," *Armed Forces Journal*, October, 118-119 (1987).

[11] National Security and International Affairs, "Need for a Lessons-Learned System at the National Training Center." Report No. GAO/NSIAD-86-130. U.S. General Accounting Office, Washington, D.C. (1986).

[12] L. Ingber, "Statistical Mechanics of Neocortical Interactions: EEG Dispersion Relations." *IEEE Trans. Biomed. Eng.* 32, 91-94 (1985).

[13] L. Ingber, "Mesoscales in Neocortex and in Command, Control and Communications (C^3) Systems," in *Systems with Learning and Memory Abilities: Proceedings, University of Paris 15-19 June 1987*, J. Delacour and J.C.S. Levy, ed., Elsevier, Amsterdam (1988), pp. 387-409.

[14] L. Ingber, "Statistical Mechanics Algorithm for Response to Targets (SMART)," in *Workshop on Uncertainty and Probability in Artificial Intelligence*: UC Los Angeles, 14-16 August 1985, AAAI-RCA, Menlo Park, CA (1985), pp. 258-264.

[15] L. Ingber, "Nonlinear Nonequilibrium Statistical Mechanics Approach to C^3 Systems," in *9th MIT/ONR Workshop on C^3 Systems: Naval Postgraduate School*, Monterey, CA, 2-5 June 1986, MIT, Cambridge, MA (1986), pp. 237-24.

[16] L. Ingber, "C^3 Decision Aids: Statistical Mechanics Application of Biological Intelligence," in 1987 *Symposium on C^3 Research: National Defense University*, Washington, D.C., 16-18 June 1987, National Defense University, Washington, D.C. (1987) pp. 49-57.

[17] L. Ingber, "Mathematical Comparison of Computer Models to Exercise Data: Comparison of JANUS(T) to National Training Center Data," in *1988 JDL C^2 Symposium: Naval Postgraduate School*, Monterey, CA, 7-9 June 1988, SAIC, McLean, VA (1988), pp. 541-549.

[18] L. Ingber, H. Fujio, and M.F. Wehner, "Mathematical Comparison of Combat Computer Models to Exercise Data." *Mathl. Comput. Modelling* submitted (1989).

[19] L. Zadeh, "A Computational Theory of Dispositions," *Int. J. Intelligent Sys.* 2, 39-63 (1987).

[20] I. R. Goodman, "A Probabilistic/Possibilistic Approach to Modeling C^3 Systems Part II," in 1987 *Symposium on C^3 Research: National Defense University*, Washington, D.C., 16-18 June 1987, National Defense University, Washington, D.C. (1988), pp. 41-48.

[21] M. Goldsmith and J. Hodges, "Applying the National Training Center Experience: Tactical Reconnaissance." Report No. N-2628-A. RAND, Santa Monica, CA (1987).

[22] L. Ingber, "Very Fast Simulated Re-annealing." *Mathl. Comput. Modelling* to be published (1989).

[23] H. Szu and R. Hartley, "Fast Simulated Annealing." *Phys. Lett.* A 122, 157-162 (1987).

[24] S. Kirkpatrick, C.D. Gelatt, Jr., and M.P. Vecchi, "Optimization by Simulated Annealing." *Science* 220, 671-680 (1983).

[25] M.F. Wehner and W.G. Wolfer, "Numerical Evaluation of Path Integral Solutions to Fokker-Planck Equations." III. Time and Functionally Dependent Coefficients. *Phys. Rev.* A 35, 1795-1801 (1987).

About the Authors

Dr. Jeffrey Abram is Manager of the Battle Management program at Advanced Decision Systems. He has been the project leader of the AirLand Battle Management project since 1986, as well as one of the principal designers and developers of the system. Dr. Abram received a B.S. degree in applied mathematics and computer science (1975), an M.S. degree in systems science and mathematics (1976), and a D.Sc. degree in systems science and mathematics (1981), all from Washington University.

Edward C. Brady is Senior Vice President and General Manager of the MITRE Corporation's Washington C³I Division. He has served on the Army Science Board and committees and task forces of the Defense Science Board and Naval Research Advisory Committee. Mr. Brady is a member of the Board of the Military Operations Research Society and Association of Unmanned Vehicle Systems. He received his B.S. degree from the U. S. Naval Academy and M.S. degrees from Georgetown University's Foreign Service Program (1976) and American University's Management Program (1985). He has completed his course work for a doctorate in economics at Georgetown University.

Dr. Rex V. Brown is Chairman of the Board of Decision Science Consortium, Inc. and has combined academic and consulting careers in the areas of decision analysis, risk analysis, statistics, and probability modeling. He has taught and researched in multiattribute decision analysis, managerial economics and statistics. He has B.A. (1957) and M.A. (1960) degrees from Cambridge University and a D.B.A. (1968) from the Harvard Business School. Dr. Brown's books include *Decision Analysis for the Manager*, *Research and the Credibility of Estimates*, and *Marketing Research and Information Systems*.

Dr. John P. Crecine is President of the Georgia Institute of Technology and a former senior vice president for Academic Affairs at Carnegie-Mellon University. He earned a bachelor's (1961), master's (1963), and doctoral degrees (1966) in industrial administration from Carnegie-Mellon University. He has served as a full professor at the University of Michigan; as an economist with the U.S. Department of Commerce; a consultant with the U.S. Bureau of Budget and an economist with The Rand Corporation. Before joining Georgia Tech, he was Chief Officer of the Inter-university Consortium for Educational Computing. He has authored three books and edited several books and monographs related to governmental budgeting and public policy. Dr. Crecine is a Fellow Commoner and Visiting Scholar at Cambridge University.

Dr. John T. Egan is on the division staff of the Tactical Electronic Warfare Division of the Naval Research Laboratory, where he is engaged in long-range EW planning. His research interests lie in information processing as it supports battle management. Dr. Egan received his Ph.D. in physics from the State University of New York, and has completed postdoctoral assignments in physics and computer science at Stanford University and the NASA Ames Research Center. He has also taught computer science within the SUNY system and at George Mason University.

Dr. Elliot E. Entin is a member of the technical staff of ALPHATECH, Inc. and is Project Leader on modeling the cognitive process under stress and investigating decision making in C³ systems. He was Associate Professor of Psychology at Ohio University. Dr. Entin received his B.S. (1962) degree from Bradley University, and his M.S. (1965) and Ph.D. (1968) degrees in education and psychology from the University of Michigan.

Dr. Paul E. Girard is employed as a Senior Scientist with the COMSYSTEMS Division of SAIC, in San Diego, California, where he is engaged in command and control system analysis, modeling, and design. He graduated from the U.S. Naval Academy in 1965 with a B.S. degree in naval architecture. He received a Ph.D. in electronic engineering (1974) from the Naval Postgraduate School and served in the Naval Undersea Center, Naval Ocean Systems Center and Naval Electronic Systems Command. While in the Office of Naval Research, Dr. Girard was instrumental in the establishment of the Joint Directors of Laboratories C^2 Research Program.

Robert W. Grayson is employed with the Washington C^3I Division of The MITRE Corporation and currently chairs the Command Center Working Group on the SDI Phase I Engineering Team. He has extensive experience in analysis of strategic forces and C^3 systems. For six years he was Director of the Air Force SABER LANCE series of annual strategic force posture studies and for five years, led the annual DCA/MITRE analyses for the Minimum Essential Emergency Communications Network (MEECN) Master Plan and MEECN Communications Plan.

Dr. Lester Ingber is Professor of Physics at the Naval Postgraduate School, Monterey, California, consultant to the U.S. Army Concepts Analysis Agency, Bethesda, Maryland, and Senior Research Associate at the Naval Ocean Systems Center, San Diego, California. He has developed a methodology for mathematical comparisons of computer combat simulations with exercise data. Dr. Ingber has published papers in atomic, nuclear, astro- and elementary particle physics. He received his B.S. degree (1962) in physics from Caltech and his Ph.D. degree (1965) in theoretical nuclear physics from the University of California, San Diego.

Dr. Stuart E. Johnson is the Director of the Command and Control Research Program, Strategic Concepts Development Center, Institute for National Strategic Studies, at the National Defense University. He served as Director, Systems Analysis in NATO Headquarters from 1982 to 1985 and as Systems Analyst in OSD/PA&E from 1979 to 1982. Dr. Johnson was a systems analyst with the Institute for Defense Analysis and with

R&D Associates. He holds a B.A. degree from Amherst College (1966) and a Ph.D. degree from the Massachusetts Institute of Technology (1971).

Dr. Thomas Kurien is a principal engineer with ALPHATECH, Inc. He is currently designing multisensor multitarget tracking algorithms using advanced concepts in estimation theory. Dr. Kurien is a member of IEEE and has authored more than 30 technical papers and reports. He periodically teaches a short course on advanced multisensor multitarget tracking at the University of California in Los Angeles. Dr. Kurien received his B.S. (1973) degree from Osmania University (India), and his M.S. (1976) and Ph.D. (1983) degrees from the University of Connecticut, all in electrical engineering.

Dr. Alexander H. Levis is Senior Research Scientist at the Massachusetts Institute of Technology Laboratory for Information and Decision Systems. He was educated at MIT where he received B.S. (1965), M.S. (1965), M.E. (1967), and Sc.D. (1968) degrees in mechanical engineering. He also attended Ripon College where he received an A.B. degree (1963) in mathematics and physics. Dr. Levis taught systems and control theory at the Polytechnic Institute of Brooklyn from 1968 to 1973 and, for the following six years, he was with Systems Control, Inc. of Palo Alto, California. Dr. Levis' research interests include command and control with emphasis on organization theory and measures of effectiveness. He has been active in professional organizations and has served as editor or associate editor for a number of journals.

Donald S. Lindberg is with the American Technologies Corporation, Fairfax, Virginia, where he manages avionics development and engineering projects in the fields of electronic warfare and airborne signal intelligence. He has operational experience in major military command staffs, fleet electronic warfare, ASW, and operational intelligence. He was a Naval Aviator and Aeronautical Engineering Officer. Mr. Lindberg is a graduate of the U.S. Naval Academy (1943) and has performed graduate studies in physics and engineering at the University of Maryland.

Dr. Jean MacMillan is a cognitive psychologist with ALPHATECH, Inc., studying human-computer interactions in command and control systems. Her most recent work involves experiments to determine the most effective role for human decision making in a strategic defense system. She received her doctorate from Harvard University for research linking human categorization decisions to adaptive network models for machine learning.

Dr. Michael F. O'Connor is a senior principal scientist at Decision Science Consortium, Inc., in Reston, Virginia. He has over 20 years experience in the fields of decision analysis, systems planning and evaluation, applied statistical analysis, applied measurement theory, and psychological research on human decision making. Dr. O'Connor has served as systems engineer, decision analyst, research associate, assistant professor, and management consultant. Much of his work has been on command and control systems, with recent emphasis on evaluation of battle management/command, control, and communications (BM/C³) architectures for the U.S. Army Strategic Defense Command. He received his Ph.D. in mathematical psychology from the University of Michigan.

Didier M. Perdu is employed by THOMSON-CSF, Inc. He is now a visiting scientist at the Massachusetts Institute of Technology Laboratory for Information and Decision Systems, with specific research interests in decision aids (especially expert systems), design and evaluation of distributed intelligent systems. Mr. Perdu was born in Paris, France, in 1962 and graduated from the École Supérieure d'Électricité (1986) and received the M.S. degree in technology and policy from the Department of Electrical Engineering and Computer Science of MIT (1988).

Daniel Serfaty is Manned Systems Group Leader at ALPHATECH, Inc. where he leads research projects in human decision making and a task to develop a theory of wartime headquarters effectiveness. He received his B.S. (1977) and M.S. (1980) degrees in Aeronautical Engineering from Technion-Israel Institute of Technology, and his M.B.A. (1984) degree from the University of Connecticut. Mr.

Serfaty is currently a Ph.D. candidate in electrical engineering and computer science at the University of Connecticut. He is a member of the IEEE and AIAA.

Dr. Harold W. Sorenson is Professor of Engineering Sciences, University of California, San Diego. He was Chief Scientist of the U.S. Air Force from 1985 to 1988 and now serves as a member of the Air Force Scientific Advisory Board. He is cofounder and former president and CEO of Orincon Corporation of La Jolla, California. Dr. Sorenson received his B.S. degree (1957) from Iowa State, and his M.S. (1963) and Ph.D. (1966) degrees from UCLA. He is a member of IEEE, a former member of the Board of Directors, IEEE, and a member of the Board of Governors, IEEE Control Systems Society. Dr. Sorenson is the author of the graduate level textbook *Parameter Estimation*.

Gregory Stachnick is a computer scientist at Advanced Decision Systems. He is the technical leader of the AirLand Battle Management project, and is the principal designer of the Maneuver Planning Expert System. He received B.A. (1969) and M.A. (1971) degrees in mathematics from San Francisco State University.

Dr. Stuart H. Starr is the Director of Plans, Washington C³I Division, The MITRE Corporation. He has served as Assistant Vice President for C³I Systems, M/A-COM Government Systems, Inc.; Director of Long-Range Planning and Systems Evaluation, Office of the Assistant Secretary of Defense (C³I); and Senior Project Leader, Institute for Defense Analysis. Dr. Starr received his B.S.E.E. and B.A. (1963) degrees in a concurrent program from Columbia University and Queens College, and his M.S. (1965) and Ph.D. (1969) degrees from the University of Illinois.

Dr. Robert R. Tenney is Vice President of Marketing, Research and Development at ALPHATECH, Inc. He taught courses in signal processing, computer architectures, modern control, optimization theory, and large-scale systems at MIT and worked in antisubmarine warfare target acquisition and tracking at the Naval Surface Weapons Center. Dr. Tenney received his B.S.E.E., M.S.E.E. (both 1976), and Ph.D. (1979) degrees from MIT.

Vice Admiral Jerry O. Tuttle, USN, is the Director, Space, Command and Control in the Office of the Chief of Naval Operations. When this paper was presented he was Director, Command, Control and Communications Systems (J-6), The Joint Chiefs of Staff. He enlisted in the Navy in 1955 and was commissioned and designated a Naval Aviator in 1956. Admiral Tuttle commanded the carrier John F. Kennedy, Carrier Group EIGHT and Carrier Group TWO/Battle Force SIXTH Fleet. He received a communications engineering degree from the Naval Postgraduate School (1963) and a master's degree in international relations from George Washington University (1969).

Dr. Harry L. Van Trees is Distinguished Professor of Information Technology, Electrical and Systems Engineering, George Mason University, Fairfax, Virginia. He was President, M/A-COM Government Systems, Inc.; Assistant Secretary of Defense (C³I); Chief Scientist, U. S. Air Force; Assistant Vice President, Communications Satellite Corporation; Chief Scientist and Associate Director of the Defense Communications Agency; and Professor of Electrical Engineering, Massachusetts Institute of Technology. He is the author of a three-volume set of books on Detection, Estimation, and Modulation Theory, one of the set now in its 23rd printing. He is a Fellow, IEEE, and recipient of the AFCEA Gold Medal for Engineering in 1988. Dr. Van Trees is a graduate of the U.S. Military Academy, received his M.Sc.E.E. degree from the University of Maryland and his Sc.D.E.E. degree from MIT.

Joseph G. Wohl (1927-1989) was a Senior Vice President of ALPHATECH, Inc. He was well-known for his many contributions to human decision making research in military applications, including the widely used SHOR paradigm for command and control decision making. He served as Vice President of the IEEE Systems, Man, and Cybernetics Society from 1980 to 1985, and was Chairman of the 1989 SMC Conference. In 1988, he received the Norbert Wiener Prize from IEEE.

Ronald E. Wright is Chief Engineer of the New Ventures Department of Ferranti International, where he is concerned with the application of new technologies to command and control systems. Prior to this, he was responsible for the development of a modular family of equipments which form the basis of many of the command and weapon systems currently in service with the British Royal Navy. Mr. Wright holds an M.Sc. degree from London University and is a Fellow of the I.E.E.